Drag Reduction of Complex Mixtures

Drag Reduction of Complex Mixtures

Keizo Watanabe
Minami-Osawa, Hachioji, Japan

ELSEVIER

ACADEMIC PRESS
An imprint of Elsevier

Academic Press is an imprint of Elsevier
125 London Wall, London EC2Y 5AS, United Kingdom
525 B Street, Suite 1650, San Diego, CA 92101, United States
50 Hampshire Street, 5th Floor, Cambridge, MA 02139, United States
The Boulevard, Langford Lane, Kidlington, Oxford OX5 1GB, United Kingdom

Notices
Knowledge and best practice in this field are constantly changing. As new research and experience broaden our understanding, changes in research methods, professional practices, or medical treatment may become necessary.

Practitioners and researchers must always rely on their own experience and knowledge in evaluating and using any information, methods, compounds, or experiments described herein. In using such information or methods they should be mindful of their own safety and the safety of others, including parties for whom they have a professional responsibility.

To the fullest extent of the law, neither the Publisher nor the authors, contributors, or editors, assume any liability for any injury and/or damage to persons or property as a matter of products liability, negligence or otherwise, or from any use or operation of any methods, products, instructions, or ideas contained in the material herein.

Library of Congress Cataloging-in-Publication Data
A catalog record for this book is available from the Library of Congress

British Library Cataloguing-in-Publication Data
A catalogue record for this book is available from the British Library

ISBN: 978-0-12-809920-9

Publisher: Matthew Deans
Acquisition Editor: Brian Guerin
Editorial Project Manager: Joshua Bayliss
Production Project Manager: Anitha Sivaraj
Designer: Mark Roger

Typeset by Thomson Digital

CONTENTS

Preface *vii*

1. Basic Concept of Complex Mixtures

1.1 Introduction 1
1.2 Classification of fluids 1
1.3 Flow property of complex mixtures 4
References 12

2. Conservation of Mass and Momentum

2.1 Introduction 13
2.2 Equation of continuity 13
2.3 Equation of fluid motion 15
2.4 Navier–stokes equation 17
2.5 Boundary conditions 20
2.6 Power-law fluids 27
References 34

3. Drag Reduction in a Heterogeneous Flow

3.1 Introduction 35
3.2 Liquid–solid two-phase flow 36
3.3 Gas–solid two-phase flow 57
3.4 Gas–liquid two-phase flow 66
References 74

4. Drag Reduction in a Pseudo-Homogeneous Flow

4.1 Introduction 75
4.2 Biofiber suspensions 76
4.3 Liquid–solid particle suspensions 81
4.4 Measurements for lift acting on a sphere 114
References 121

5. Drag Reduction in a Homogeneous Flow

5.1 Introduction 123
5.2 Polymer solutions 124
5.3 Surfactant solutions 157
5.4 Biopolymer solutions 194
References 203
Index 205

PREFACE

From a historical point of view, the hydraulic machine developed to provide a method for advantageously obtains water, which is the necessity for human survival, and it encouraged development of the fluid mechanics. At present, many kinds of flow phenomenon have been analyzed by the numerical solution of the computer using the mathematical modeling and formulation for those flows. However, the flow phenomena have the property of being multiple and it is complicated, about the turbulent flow, many questions still remain unclear.

On the other hand, since the analysis of drag in a fluid flow is closely related to improve the performance of hydraulic machine or the evaluation of the transportation power for the pipeline systems, the calculation would be a particularly important issue in the fluid engineering. In addition, the objective fluid is expanded not only to a single phase fluid, but also two phase fluid, and we recognize an increasing complexity of flow behavior.

It is well known that the fluid transport systems are used in many industrial fields or surrounding living environment since a fluid can transport energy. Thus, it is very important to reduce the transportation power and we will expect to apply the energy saving system for preventing global warming if it is accomplished by hydrodynamic knowledge. A word to the wise, we need to grasp the flow phenomenon in order to make it come true.

This book is written in order to clear the specific property of drag reduction phenomenon in the flow of complex mixtures by the experimental results under such a background.

For fluid including solid or gas phase that it is bite-size, we can assume a continuum model to the fluid by taking no notice the effect of the size on the flow behavior. But, it cannot be ignored according to increase of the size. In this book, the drag reduction phenomenon is divided into three categories that they are heterogeneous, pseudo-homogenous and homogenous flow by pay attention a particle size in fluids. In addition, the flow field is intended not only for a pipe flow but also for an external flow that it is flow around a sphere or a circular cylinder, and the visualization photographs are shown to cover as wide a range of subject as possible. The study on the drag reduction will give someone a clue of the clarification of a turbulent flow, which has many unexplained problems in fluid mechanics.

Chapter 1 presents the basic concept of complex mixtures, including an introductory particle size in fluid. Because the pressure loss or drag decreases from the standard value in the drag reduction phenomenon, the basic knowledge is necessary to analysis it. Equations for conservation of mass and momentum, which is dealt in the fluid mechanics, are reviewed in Chapter 2. It will be necessary to consider the physical significance for the deviation or the change including the assumption or the condition setting to the flow fields. The drag reduction phenomenon of complex mixtures is divided into three flow categories that it is called as a heterogeneous, a pseudo-homogeneous and a homogeneous flow to grasp an image, which makes the contents of all or a portion on the phenomenon. A heterogeneous flow that the most of two-phase flows may become a target is written in Chapter 3. Although it is applied in the hydraulic or the pneumatic transportation systems and many experimental data on the pressure drop in a circular pipe have been published, the difficulty remains in a systematic formulation of the data because of the complexity in flow behavior. Thus the drag reduction phenomenon, which is described in this chapter, may be limited.

In Chapter 4, the drag reduction in the flow of a pseudo-homogeneous fluid is described. It is the micron order in the size of particle and it is called suspensions frequently. There are a wide range of an included particle. Since the flow curve is obtained according to the terms, we can decide the apparent viscosity of the suspensions. On the other hand, an effect of immigration of particles in liquid on the separation is clarified experimentally.

Drag reduction of polymer or surfactant solutions, which becomes the target to primarily drag reduction study, are summarized in Chapter 5. However, there are few studies on an external flow of surfactant solutions. The deference of phenomenon between polymer and surfactant solutions is shown by the flow pattern of an external flow. It may be become an important issue to consider the analogy on the phenomenon between the pseudo-homogeneous and the homogeneous flow in future.

I hope this book will help more researchers or students become interested in the drag reduction phenomenon.

Many sources for the experimental results in this book have been drawn from papers that were produced at the Fluid Engineering Laboratory in the Faculty of Engineering at Tokyo Metropolitan University.

The author express the appreciation to my teacher Hiroshi Kato, a professor emeritus at Tokyo Metropolitan University fostered my interest in drag reduction of polymer solutions to those who made the writing of this book possible. The author also acknowledges gratefully the assistance and cooperation of Dr. Satoshi Ogata and the student of Tokyo Metropolitan University at the time.

Keizo Watanabe

CHAPTER 1

Basic Concept of Complex Mixtures

Contents

1.1 Introduction 1
1.2 Classification of Fluids 1
1.3 Flow Property of Complex Mixtures 4
References 12

1.1 INTRODUCTION

All material existing on the earth are classified into three categories of solid, liquid, and gas. Since liquid or gas material flow when an external force acts on these materials, they are called as fluids. Water or air is something very familiar and important fluid for us because we could not live without them. By thinking about the method for obtaining water without difficulty, it promoted the development from a well bucket to hydraulic machine. In parallel with such processing, hydraulics and fluid mechanics developed in order to grasp the physical properties of flow. We can understand from historical fact that ancient civilizations were born along the banks of great waterways such as the Nile and the Yellow River.

On the other hand, a flow of gas becomes familiar to us from information on windmill, airplane or rocket. And also, we know debris flow or mudflow caused by a typhoon accompanied by heavy rain through the disaster information is acquired from local TV broadcasting. In this way, even the fluids flow are covered including water flow, gas flow, high-speed flow, debris flow, etc., with extremely large variety.

In this chapter, the classification of fluids and the characteristics of flow are summarized to understand a meaning or a situation for the flow phenomenon.

1.2 CLASSIFICATION OF FLUIDS

Fluid flow can be broadly classified into two categories such as single phase and multiphase flows. Fig. 1.1 presents the classification of fluid flow. Single phase flow is a flow only for liquid or gas and the fluid state

Drag Reduction of Complex Mixtures
http://dx.doi.org/10.1016/B978-0-12-809920-9.00001-9

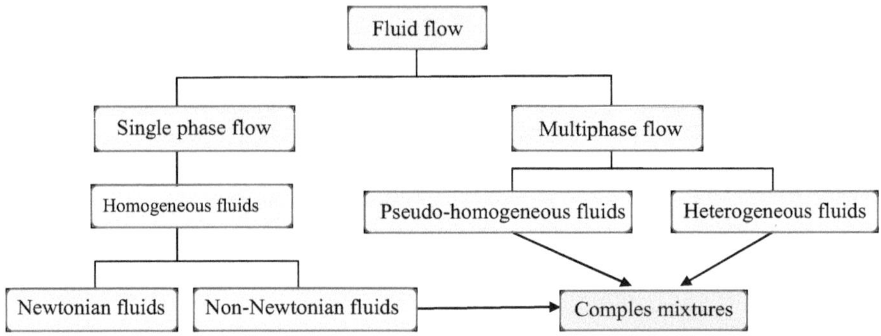

Figure 1.1 ***Classification of fluids.***

is homogenous from the viewpoint of continuum hydrodynamics. And also, these fluids are divided into Newtonian and non-Newtonian fluids from the viewpoint of the viscous behavior. Water or glycerin solutions are Newtonian fluids which satisfy Newton's law of viscosity. A kind of high molecular polymer, bio-polymer or surfactant solutions is called as non-Newtonian fluids that has a nonlinear characteristic in the viscosity or a visco-elastic property.

Multiphase flow is comprised of solid particles or small bubbles in fluids, and they are classified into four types which are gas–liquid, solid–liquid, solid–gas, and liquid–liquid mixtures flow. Although the behavior of single phase flow might be able to analyze from the viewpoint of true homogeneous flow, that of the multiphase mixtures will be treated as the heterogeneous mixtures flow according to the increase in particle size. The extensive and widely experimental data relating to the flow in pipes of non-Newtonian fluids and other complex mixtures are summarized by Govier and Aziz[1]. In this book, non-Newtonian fluid and multiphase fluids are collectively referred to as complex mixtures as well as them. As described above, the term of complex mixtures is used a general expression referring for non-Newtonian fluids, mixing liquid/solid or gas/solid multiphase stream in this book.

In general, the flow characteristics largely depend on the size of particle contained in fluids. Fig. 1.2 indicates a typical example of particle size of the matters. The size is only a guide, and it is an important to know the determination method for the characteristics length of an actually measured particle. In the figure, the ranges of Brownian motion and the visible rays are presented for a comparison of these particle sizes. Generally speaking, we might be able to treat as a pseudo-homogenous flow in the case of liquid–solid two phase fluids with fine solid particles less than 50 μm in

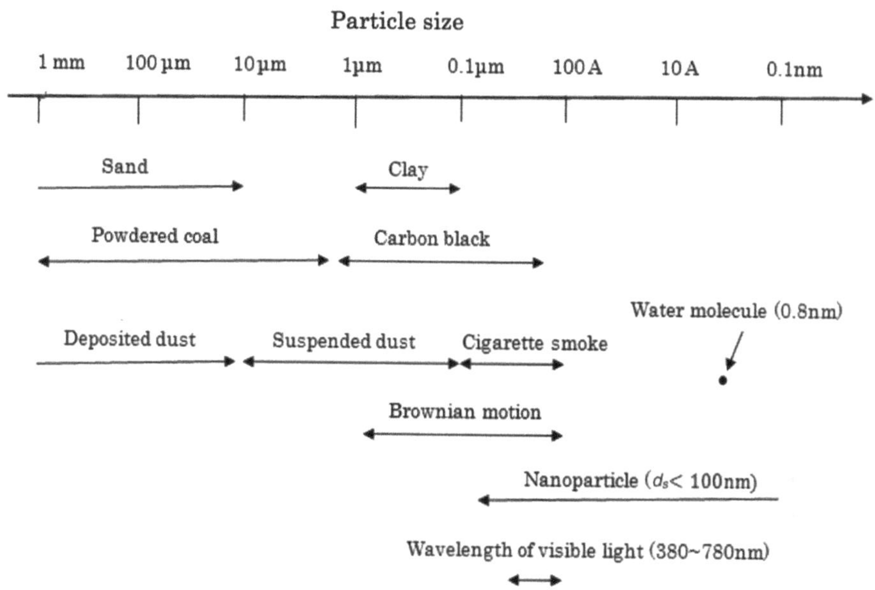

Figure 1.2 ***Typical example of particle size.***

a diameter. However, it will be a meaningful discussion that we grasp to the flow behavior from a microscopic viewpoint for the existence in fluids because the additives have different shapes. By using these additives, the characteristics of mixtures changes in the internal physical structure or the physical property.

One of the notable problems in engineering fields is the change of drag or pressure loss in the flow characteristics. As well known, the flow regime of real fluids is classified as laminar or turbulent flow on the basis of flow structure. In general, laminar is a smooth flow without turbulence and turbulent is a random flow with three-dimensional structure turbulence. As a matter of course, entirely different mechanisms on the flow drag are responsible for such laminar or turbulent flow. By adding certain material into liquid or gas, the turbulence structure will be changed in comparison with the state before it.

On the other hand, many flow fields exist in our daily experience. They may be divided first into an internal and an external flow. Former is a flow in a circular pipe or a duct, and the latter is a flow around a blunt body that it is a sphere or a circular cylinder. As presented in Fig. 1.3, when the complex mixtures in which certain matters were added in fluids flow in a pipe, it is an example of an internal flow, the pressure drop of pipe will change under its condition. The total pressure loss Δp_t will be given as follows,

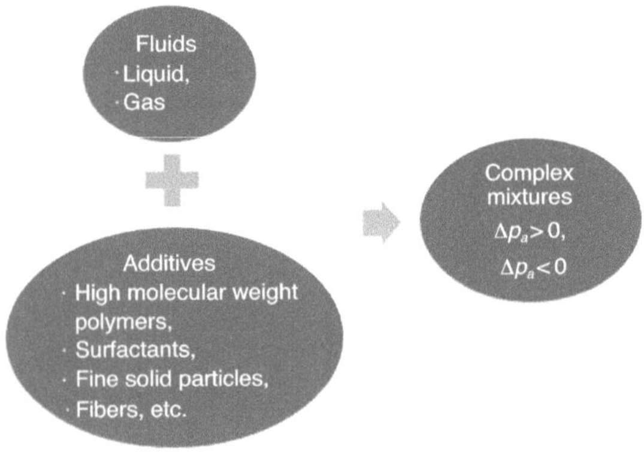

Figure 1.3 ***Effect of an added or a mixed matter on the pressure loss.***

$$\Delta p_t = \Delta p_f + \Delta p_a \tag{1.1}$$

where Δp_f and Δp_a are the pressure loss only fluids and the additional pressure, respectively. In the case of heterogeneous fluids flow, Δp_a in Eq. (1.1) becomes the term with a (+). However, the term with a (−) exists in the case of fluids added high molecular weight polymers or surfactants. It is obtained also in an external flow. Thus, this phenomenon is called "Drag reduction" as a general term. It will be a major advantage effect since the effect relates directly to save the energy of hydraulic transport systems. In addition, we can discuss the turbulent structure by analyzing the deviate from normal situation.

1.3 FLOW PROPERTY OF COMPLEX MIXTURES

As described above, the drag reduction is one of the major advantage property of the complex mixtures flow. In general, the phenomenon is classified into two categories, such as a passive control and an active control technique. The latter is a flow system with energy supply from the outside of the system, and the former is the system without energy supply. Although a passive control technique does not continuously need energy during the operation and a large drag reduction ratio is obtained, the application is limited to a closed system since the degradation occurs or the environmental load increases. On the other hand, an active control technique may be used in the case that the cost of the technique exceeds than that of the operating additional cost when it is used.

Fig. 1.4 represents the classification of drag reduction obtained by passive control technique. Furthermore, as shown in the figure, the technique is classified into two categories of drag reducing additives and drag reducing walls from the consideration that the drag is an interaction between fluids and the wall surface. In both cases, the target is a turbulent modification in turbulent flow.

In 1948, B. A. Toms[2] reported the phenomenon that pressure loss of a pipe flow in turbulent flow range decreases in comparison with that of Newtonian fluid flow by addition of a small amount of polymer into solvent. It is called as "Toms effect", and many studies have been done for it until now. Although polymer or surfactant solutions may be treated as a homogenous flow, solid particle suspensions become pseudo-homogenous or heterogeneous mixtures in some cases.

Riblets,[3] which are microgrooves on the surface, will be an effective drag reduction technique for both liquid or gas flows. We can obtain skin friction drag reduction in the range of 5%–8% in maximum. For laminar flow, the application of a hydrophobic wall with fluid slip[4] may be the only drag reduction technique since the turbulence does not exist in the flow range.

The drag reduction phenomenon of heterogeneous or pseudo-homogenous fluids depends on some nondimensional parameters based on a characteristic length of a particle. For the calculation and the estimation of the parameters, the difficulty remains in the determination of the characteristics

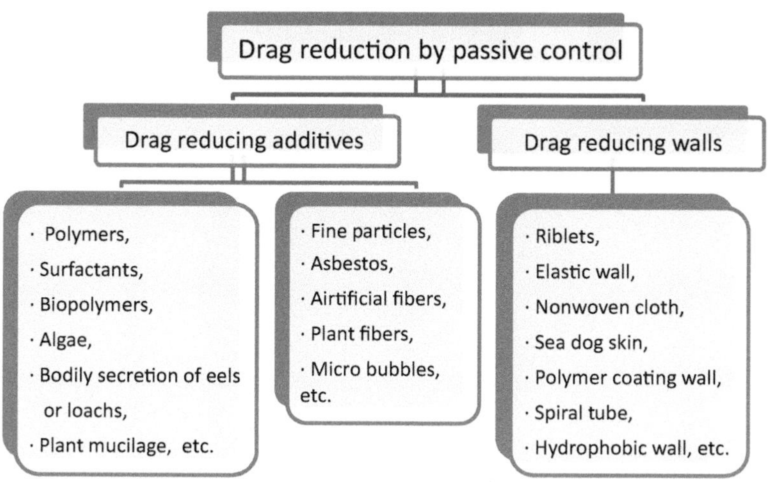

Figure 1.4 ***Classification of drag reduction.***

length of including particles since the shape that was very different not a sphere simply. In generally, we use a mean diameter for the characteristics of length of a particle. There are two methods for obtaining particle size from a direct view or the microscope and a measuring for the terminal or the floating velocity. The latter method might be in agreement with the characteristics length of particle in complex mixtures flow. However, when we analyze the behavior of fibers or nonspherical particles in fluids from viewpoint of Lagrange's method, it will be necessary to know the values of an aspect ratio and Yang's modules for the calculation. It enhances the importance of certain model parameter for the simulation of the flow behavior.

Since a drag of external flows is given as the sum of skin friction and form drag and the ratio differs from flow range to flow range, it is very important for an evaluation of the drag to grasp the flow behavior. Thus, data on velocity profile and pressure distribution will be necessary for the calculation of the drag.

Fig. 1.5 represents the correlation of the information obtained each condition. We can obtain the physical constants of mixed matter, solvent

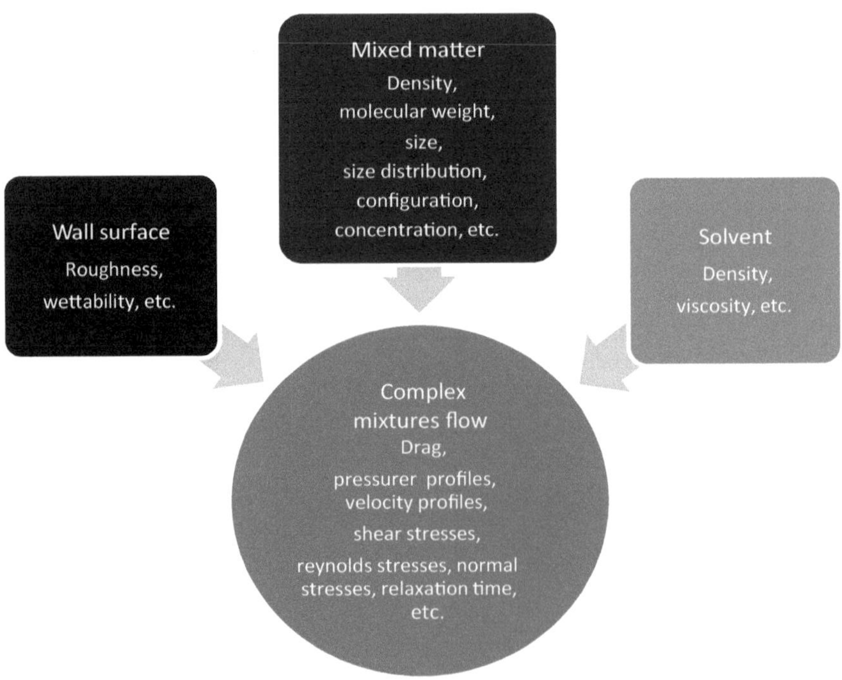

Figure 1.5 ***Correlation chart of physical quantity.***

and wall surface at stage before the mixture flow. As described above, the data on configuration or size of the matter are important to consider the flowing conditions. In the flow state, the pressure and velocity distributions will be effective experimental data in order to examine the drag. Although some measuring methods for the physical constant of complex mixtures are established, it is difficult to say at present that the measurement system on physical constants related the visco-elastic effect has been established. That will probably be a future problem in order to systematize the quantitative evaluation for the drag reduction phenomenon. From the viewpoint of industrial usefulness, the effect is required having long lasting on the occasion of the application. Such as, the degradation does not occur.

Because the flow of complex mixtures is very complicate, the dimensional analysis may be a significant method for the analysis of drag when we cannot analyze it by the equation of motion. It is analyzed by using Buckingham π theorem described as below.
When the physical phenomenon is affected by n physical constants, $f(A_1, A_2, A_3 \cdots\cdots\cdots A_n) = 0$, if the number of the fundamental unit in the system is m, we can express the phenomenon by $(n - m)$ nondimensional parameters.

We may express mathematically the relationship among the variables function form,

$$f(A_1, A_2, A_3 \cdots\cdots\cdots A_n) = 0 \tag{1.2}$$

Eq. (1.2) can be expressed as:

$$F(\pi_1, \pi_2, \pi_3 \cdots\cdots \pi_{n-m}) = 0 \tag{1.3}$$

or

$$\pi_1 = G(\pi_2, \pi_3 \ldots\ldots \pi_{n-m}) \tag{1.4}$$

$\pi_1, \pi_2, \pi_3 \cdots\cdots \pi_{n-m}$ in Eq. (1.3) are the nondimensional parameters given as the ratio the product of the physical constants of $A_1, A_2, A_3 \cdots\cdots A_m$ and $A_{m+1}, A_{m+2}, \cdots\cdots A_n$.

In other words, the physical constants, $A_{m+1}, A_{m+2}, \cdots\cdots A_n$ is given by using the nondimensional parameters, $\pi_1, \pi_2, \pi_3 \cdots\cdots \pi_{n-m}$ as follows,

$$A_{m+1} = \pi_1 A_1^{x_1} A_2^{y_1} A_3^{r_1} \cdots\cdots A_m^{z_1}$$
$$A_{m+2} = \pi_2 A_1^{x_1} A_2^{y_2} A_3^{r_2} \cdots\cdots A_m^{z_2}$$
$$A_n = \pi_{n-m} A_1^{x_{n-m}} A_2^{y_{n-m}} A_3^{r_{n-m}} \cdots\cdots A_m^{z_{n-m}} \tag{1.5}$$

where, the exponent groups, $(x_1, y_1, r_1 \cdots\cdots z_1,\ x_2, y_2, r_2 \cdots\cdots z_2,\ x_{n-m}, y_{n-m}, r_{n-m} \cdots\cdots z_{n-m})$ are decided by the condition that the dimension of both members of Eq. (1.5) must be equal.

A group of $A_1, A_2, A_3 \cdots\cdots A_m$ is a basic physical constant, and the selection is important although it is $m = 3$, for example, (M, L, T) in the number of primary dimension for an isothermal system. In the selection, we should not include the physical constant to be analyzed in the basic physical constant.

As shown in Fig. 1.5, there are many physical constants which affects to the flow behavior, thus it may be necessary to set up the physical constant by the selecting conditions set. As a matter of course, the functional relationship among the nondimensional parameter, Eq. (1.4), must be determined experimentally.

Example 1.1 Pressure loss of solid–liquid two-phase flow in a circular pipe.

As noted in Fig. 1.5, the pressure loss of solid–liquid two-phase flow Δp, depends on the liquid mean velocity U_f, the liquid density ρ_f, the viscosity μ, the pipe diameter D, the pipe length L, the pipe roughness k, and then, it depends on the velocity U_s, the diameter d_s, and the density ρ_s the solid particle, respectively. Determine the set of nondimensional group that affect to the pressure loss.

Solution

This phenomenon can be written as follows,

$$f\left(\Delta p_t, U_f, \rho_f, \mu, U_s, d_s, \rho_s, D, k, L\right) = 0 \tag{1.6}$$

At first, if we select the liquid mean velocity U_f, the liquid density ρ_f, and the pipe diameter D as the basic physical constant, Eq. (1.6) will be given as,

$$\pi_1 = F\left(\pi_2, \pi_3, \pi_4, \pi_5, \pi_6\right) \tag{1.7}$$

where $\pi_1 \sim \pi_6$ are the nondimensional parameter as:

$$\Delta p_f = \pi_1 \rho_f^{x_1} U_f^{y_1} D^{z_1}$$
$$\mu = \pi_2 \rho_f^{x_2} U_f^{y_2} D^{z_2}$$
$$U_s = \pi_3 \rho_f^{x_3} U_f^{y_3} D^{z_3}$$
$$\rho_s = \pi_5 \rho_f^{x_5} U_f^{y_5} D^{z_5}$$
$$k = \pi_6 \rho_f^{x_6} U_f^{y_6} D^{z_6}$$
$$L = \pi_7 \rho_f^{x_7} U_f^{y_7} D^{z_7} \quad (1.8)$$

The dimensional matrix used to solve the simultaneous equations are obtained as,

	ρ_f	U_f	D	Δ_{pf}	μ	Us	d_s	ρ_s	k	L
M	1	0	0	1	0	0	0	1	0	0
L	−3	1	1	−1	1	1	1	−3	1	1
T	0	−1	0	−2	−1	−1	0	1	0	0

The unknown quantities are determined as

$$x_1 = 1,\ y_{1,} = 2,\ z_1 = 0\ ;\ \pi_1 = \left(\Delta p / \rho_f U_f^2\right) \rightarrow E = \left(\Delta p / \frac{1}{2} \rho_f U_f^2\right)$$

$$x_2 = 1,\ y_{2,} = 1,\ ;\ z_2 = 1\ \pi_2 = \left(\mu / \rho_f U_f D\right) \rightarrow R_e = \left(\rho_f U_f D / \mu\right)$$

$$x_3 = 0,\ y_{3,} = 1,\ z_3 = 0\ ;\ \pi_3 = \left(U_s / U_f\right)$$

$$x_4 = 0,\ y_{4,} = 0\ z_4 = 1\ ;\ \pi_4 = \left(d_s / D\right)$$

$$x_5 = 1,\ y_{5,} = 0,\ z_5 = 0\ ;\ \pi_5 = \left(\rho_s / \rho_f\right)$$

$$x_6 = 0,\ y_{6,} = 0,\ z_6 = 1\ ;\ \pi_6 = \left(k / D\right)$$

$$x_7 = 0,\ y_{7,} = 0,\ z_7 = 1\ ;\ \pi_7 = \left(L / D\right)$$

(Continued)

Example 1.1 Pressure loss of solid–liquid two-phase flow in a circular pipe (*cont.*)

Thus

$$\left(\frac{\Delta p}{\rho_f U_f^2}\right) = F\left(\frac{\mu}{\rho_f U_f D}, \frac{U_s}{U_f}, \frac{d_s}{D}, \frac{\rho_s}{\rho_f}, \frac{k}{D}, \frac{L}{D}\right) \tag{1.9}$$

Since the characteristics length changes in fluids engineering field, a nondimensional parameter is frequency used to arrangement the data of loss or drag. It is effective at predicting the parameter that gives rise to the phenomenon of pressure loss of the complex mixtures.

Example 1.2 Flow around a rotating disk

When a disk is rotating in viscous liquid at the constant angular velocity, the frictional moment occurs. Determine the frictional moment, M acting on the disk surface on both side.

Solution

The frictional moment depends on the disk radius, a, the angular velocity of the disk, ω, the fluid density, ρ, and the fluid viscosity, μ.It can be expressed by the physical constants, $n = 5$ as follows,

$$f(M, a, \omega, \mu, \rho) = 0 \tag{1.10}$$

Since the number of primary units, m equals 3, Eq. (1.10) becomes,

$$F(\pi_1, \pi_2) = 0 \tag{1.11}$$

If we select a. ρ. ω as the primary unit, M and μ are given as:

$$\left.\begin{aligned} M &= \pi_1 a^{x_1} \omega^{y_1} \rho^{z_1} \\ \mu &= \pi_2 a^{x_2} \omega^{y_2} \rho^{z_2} \end{aligned}\right\} \tag{1.12}$$

The dimensional matrix is,

	a	ω	ρ	M	μ
M	0	0	1	1	1
L	1	0	−3	2	−1
T	0	1	0	−2	−1

Therefore, the unknown quantities, (x_1, y_1, z_1) and (x_2, y_2, z_2) are decided by the equation:

$$\left.\begin{array}{ll} [M]: & 1 = z_1 \\ [L]: & 2 = x_1 - 3z_1 \\ [T]: & -2 = -y_1 \end{array}\right\} \quad z_1 = 1, y_1 = 2, x_1 = 5 \tag{1.13}$$

$$\left.\begin{array}{ll} [M]: & 1 = z_2 \\ [L]: & -1 = x_2 - 3z_2 \\ [T]: & -1 = -y_2 \end{array}\right\} \quad z_2 = 1, y_2 = 1, x_2 = 2 \tag{1.14}$$

The nondimensional parameter are given as:

$$\left.\begin{array}{l} \pi_1 = \dfrac{M}{\rho a^5 \omega^2} \\ \pi_2 = \dfrac{\mu}{\rho a^2 \omega} \end{array}\right\} \tag{1.15}$$

Therefore, by rearranging, π_1 and $\pi_2$in Eq. (1.15), the following equation may be obtained:

$$\left.\begin{array}{l} Cm = \dfrac{M}{\dfrac{1}{2}\rho a^5 \omega^2} \\ \mathrm{Re} = \dfrac{\rho a^2 \omega}{\mu} \end{array}\right\} \tag{1.16}$$

C_m and Re are defined as the moment coefficient and the Reynolds number, respectively.

Since Eq. (1.11) is rearranged as $\pi_1 = f(\pi_2)$, we can arrange the experimental data on the frictional resistance of a rotating disk by following equation:

$$C_m = f(\mathrm{Re}) \tag{1.17}$$

The reader can see that this result is reasonable on the frictional resistance of a rotating disk in infinity fluids that is at the rest by comparing with Fig. 5.13 in Chapter 5.

The reader may understand this method of analysis by deciphering the exercises.

Because so few real flow can be solved exactly by analytical methods alone, the dimensional analysis is a very important analytical method for fluid analysis or fluid design. By using the law, we can analytically grasp the similarity of model and prototype flow that they must be satisfied the geometric and kinematic similarity. As it was understood by solving the example relating Buckingham π theorem, we need to consider the following factor to obtain the reasonable result.

They are

- to clear the physical constant to influence the phenomenon,
- to select n unit of the basic physical constant that it is not involved the target the physical constant for analysis,
- to make the dimensional matrix and to solve the simultaneous equations, and
- to arrange the obtained parameter by considering the physical meaning.

Other example of the arrangement of the experimental data for biosuspensions obtained by the dimensional analysis is presented in Chapter 4.

REFERENCES

1. Govier GW, Aziz K. *The flow of complex mixtures in pipes*. Robert E. Krieger Publishing Co; 1972.
2. Toms BA. *Proc. First Intern. Cong. Rheology* 1949; 135 Part 2.
3. Walsh MJ, Lindemann AM. *IAA paper 84–0347*. 1984.
4. Watanabe K. Laminar drag reduction. Bentham Publishers Co; 2015. p. 11.

CHAPTER 2

Conservation of Mass and Momentum

Contents

2.1 Introduction	13
2.2 Equation of Continuity	13
2.3 Equation of Fluid Motion	15
2.4 Navier–Stokes Equation	17
2.5 Boundary Conditions	20
2.6 Power-Law Fluids	27
References	34

2.1 INTRODUCTION

In fluid mechanics, we assume that a fluid is a continuum by ignoring the discontinuity, which molecules and an atom show. It is a model having continuity, homogeneity, and isotropy. In other words, when we consider a microelement in fluid, the effects of molecular on the physical property or the flow is neglected and it is merely considered that the element is consisted by the fluid in the limit. Thus, the expression of the equation obtained by the continuity or momentum equations, become the relatively simple form. But the assumption may not be applicable for complex mixtures although it is affected by the scale or density of blend material.

Because the drag reduction ratio means the decreasing value from the standard value analyzed using the continuum model, it is significant to understand the basic equation of fluid mechanics. In this chapter, the equations of continuity and motion are summarized, and the exercises are given to understand an analytical method for basic flow problem.

2.2 EQUATION OF CONTINUITY

Any flow of complex mixtures is any subject to the law of physics, such as the conservation of mass or energy is satisfied in the flow system. Continuity equation is derived from law of conservation of mass that the total mass of matter is constant in any closed system even if any physical or chemical change occurs.

Drag Reduction of Complex Mixtures
http://dx.doi.org/10.1016/B978-0-12-809920-9.00002-0

For compressible fluid, it is given by using rectangular coordinates as follows,

$$\frac{\partial \rho}{\partial t}+\frac{\partial}{\partial x}(\rho u)+\frac{\partial}{\partial y}(\rho \upsilon)+\frac{\partial}{\partial z}(\rho w)=0 \tag{2.1}$$

where u, υ, w and ρ are the velocity components in rectangular coordinates and density of fluids respectively.

As fluid density is constant value in the case of incompressible fluid, Eq. (2.1) becomes as following equation.

$$\frac{\partial u}{\partial x}+\frac{\partial \upsilon}{\partial y}+\frac{\partial w}{\partial z}=0 \tag{2.2}$$

It should be noted that Eq. (2.1) or (2.2) is satisfied in both cases of steady or unsteady flow.

We may write Eqs. (2.1) and (2.2) more conveniently in vector symbolism.

$$\text{Compressible fluid; } \frac{\partial \rho}{\partial t}+(\nabla \cdot \rho \mathbf{V})=0 \tag{2.3}$$

$$\text{Incompressible fluid; } (\nabla \cdot \mathbf{V})=0 \tag{2.4}$$

It is necessary to use the equation of continuity as written in cylindrical coordinates for the analysis of flow in a circular pipe. When the velocity is given as $\mathbf{V}(\upsilon_r, \upsilon_\theta, \upsilon_z)$ in cylindrical coordinates and $\boldsymbol{V}\,(\upsilon_r, \upsilon_\theta, \upsilon_\varphi)$ in spherical coordinates, these equations are given as follows,

Cylindrical coordinates:

$$\text{Compressible fluid; } \frac{\partial \rho}{\partial tt}+\frac{1}{r}\frac{\partial\left(r\rho \upsilon_r\right)}{\partial rr}+\frac{1}{r}\frac{\partial\left(\rho \upsilon_\theta\right)}{\partial \theta}+\frac{\partial\left(\rho \upsilon_z\right)}{\partial z}=0 \tag{2.5}$$

$$\text{Incompressible fluid; } \frac{1}{r}\frac{\partial\left(r\upsilon_r\right)}{\partial r}+\frac{1}{r}\frac{\partial \upsilon_\theta}{\partial \theta}+\frac{\partial \upsilon_z}{\partial z}=0 \tag{2.6}$$

Spherical coordinates

Compressible fluid;

$$\frac{\partial \rho}{\partial tt}+\frac{1}{r^2}\frac{\partial}{\partial r}\left(\rho r^2 \upsilon_r\right)+\frac{1}{r\sin\theta}\frac{\partial}{\partial \theta}\left(\rho \upsilon_\theta \sin\theta\right)+\frac{1}{r\sin\theta}\frac{\partial}{\partial \varphi}\left(\rho \upsilon_\phi\right)=0 \tag{2.7}$$

Incompressible fluid;

$$\frac{1}{r^2}\frac{\partial}{\partial r}\left(r^2 \upsilon_r\right)+\frac{1}{r\sin\theta}\frac{\partial}{\partial\theta}\left(\upsilon_\theta \sin\theta\right)+\frac{1}{r\sin\theta}\frac{\partial \upsilon_\varphi}{\partial\varphi}=0 \tag{2.8}$$

2.3 EQUATION OF FLUID MOTION

From the historical viewpoint, fluid motion was analyzed by using an ideal fluid model because of its complexity. In that case, the equation of motion becomes a simple form as no shear stresses are present in an inviscid flow. However, all real fluids possess viscosity. Such as, the stresses occur in fluid by the deformation when it flows.

The equation of motion for real fluid isothermal flow can be obtained by rearranging Euler's equations. For the equation in rectangular coordinates, these expressions are

$$\rho\left(\frac{\partial u}{\partial t}+u\frac{\partial u}{\partial x}+\upsilon\frac{\partial u}{\partial y}+w\frac{\partial u}{\partial z}\right)=\rho g_x-\frac{\partial p}{\partial x}+\left(\frac{\partial \tau_{xx}}{\partial x}+\frac{\partial \tau_{yx}}{\partial y}+\frac{\partial \tau_{zx}}{\partial z}\right) \tag{2.9}$$

$$\rho\left(\frac{\partial \upsilon}{\partial t}+u\frac{\partial \upsilon}{\partial x}+\upsilon\frac{\partial \upsilon}{\partial y}+w\frac{\partial \upsilon}{\partial z}\right)=\rho g_y-\frac{\partial p}{\partial y}+\left(\frac{\partial \tau_{xy}}{\partial x}+\frac{\partial \tau_{yy}}{\partial y}+\frac{\partial \tau_{zy}}{\partial z}\right) \tag{2.10}$$

$$\rho\left(\frac{\partial w}{\partial t}+u\frac{\partial w}{\partial x}+\upsilon\frac{\partial w}{\partial y}+w\frac{\partial w}{\partial z}\right)=\rho g_z-\frac{\partial p}{\partial z}+\left(\frac{\partial \tau_{xz}}{\partial x}+\frac{\partial \tau_{yz}}{\partial y}+\frac{\partial \tau_{zz}}{\partial z}\right) \tag{2.11}$$

where the first index of τ_{ij} indicates the direction of the normal to the surface on which the stress is considered and the second index indicates the direction in which the stress acts. We can drive also these equations by using Newton's second law to that a small volume element moving with the fluid is accelerated.

Consider first the microcube of fluids that the volume is $dxdydz$ and the velocity is $V_t(x,y,z,t)$ at arbitrary position in rectangular coordinates represented (x,y,z) as shown in Fig. 2.1. If the microcube moves to $(x+\Delta x, y+\Delta y, z+\Delta z)$ after Δt, the velocity becomes $V_{t+\Delta t}(x+\Delta x, y+\Delta y, z+\Delta z, t+\Delta t)$. The acceleration, α is reduced from Taylor expansion as follows,

$$\alpha=\lim_{\Delta t\to 0}\frac{\Delta V}{\Delta t}=\lim_{\Delta t\to 0}\frac{(V_{t+\Delta t}-V_t)}{\Delta t}=\lim_{\Delta t\to 0}\left(\frac{\Delta V}{\Delta t}+\frac{\Delta V}{\Delta x}\frac{\Delta x}{\Delta t}+\frac{\Delta V}{\Delta y}\frac{\Delta y}{\Delta t}+\frac{\Delta V}{\Delta z}\frac{\Delta z}{\Delta t}\right)$$

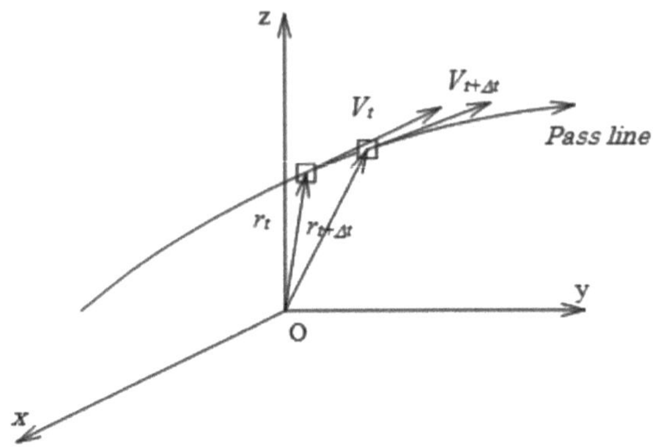

Figure 2.1 ***Velocity change of an element of fluids.***

And also, because the velocity components $V(u,\upsilon,w)$ are given as $\lim_{\Delta t\to 0}\frac{\Delta x}{\Delta t}=u$, $\lim_{\Delta t\to 0}\frac{\Delta y}{\Delta t}=\upsilon$, and $\lim_{\Delta t\to 0}\frac{\Delta z}{\Delta t}=w$, respectively, α becomes,

$$\alpha=\frac{\partial V}{\partial t}+u\frac{\partial V}{\partial x}+\upsilon\frac{\partial V}{\partial y}+w\frac{\partial V}{\partial z}$$

Also, it is represented using the operator

$$\frac{DV}{Dt}=\frac{\partial V}{\partial t}+u\frac{\partial V}{\partial x}+\upsilon\frac{\partial V}{\partial y}+w\frac{\partial V}{\partial z}$$

where $\frac{D}{Dt}$ is defined as the substantial derivative. This operator is presented by the vector representation as

$$\frac{D}{Dt}=\frac{\partial}{\partial t}+(V\cdot\nabla)$$

$(V\cdot\nabla)$ becomes in rectangular coordinates,

$$(\mathbf{V}.\nabla)=u\frac{\partial}{\partial x}+\upsilon\frac{\partial}{\partial y}+w\frac{\partial}{\partial z}$$

Terms of the left-hand side of Eqs. (2.9)–(2.11) indicate the acceleration par unit mass.

As described above, curvilinear coordinates are sometimes more convenient. By the transformation from rectangular to these coordinates, the acceleration components, $(\alpha_r,\alpha_\theta,\alpha_z)$ are express for cylindrical coordinates, (r, θ, z) as follows, where $(\upsilon_r,\upsilon_\theta,\upsilon_z)$ are the velocity components.

$$\alpha_r = \frac{\partial \upsilon_r}{\partial t} + \upsilon_r \frac{\partial \upsilon_r}{\partial r} + \frac{\upsilon_\theta}{r}\frac{\partial \upsilon_r}{\partial \theta} - \frac{\upsilon_\theta^2}{r}\upsilon_z \frac{\partial \upsilon_r}{\partial z} \tag{2.12}$$

$$\alpha_\theta = \frac{\partial \upsilon_\theta}{\partial t} + \upsilon_r + \frac{\partial \upsilon_\theta}{\partial r} + \frac{\upsilon_\theta}{r}\frac{\partial \upsilon_\theta}{\partial \theta}\frac{\upsilon_r \upsilon_\theta}{r} + \upsilon_z \frac{\partial \upsilon_\theta}{\partial z} \tag{2.13}$$

$$\alpha_z = \frac{\partial \upsilon_z}{\partial t} + \upsilon_r \frac{\partial \upsilon_z}{\partial r} + \frac{\upsilon_\theta}{r}\frac{\partial \upsilon_z}{\partial \theta} + \upsilon_z \frac{\partial \upsilon_z}{\partial z} \tag{2.14}$$

For spherical coordinates, (r,θ,ϕ), we getwhere $(\upsilon_r,\upsilon_\theta,\upsilon_\phi)$ are the velocity components.

$$\alpha_r = \frac{\partial \upsilon_r}{\partial t} + \upsilon_r \frac{\partial \upsilon_r}{\partial r} + \frac{\upsilon_\theta}{r}\frac{\partial \upsilon_r}{\partial \theta} + \frac{\upsilon_\phi}{r\sin\theta}\frac{\partial \upsilon_r}{\partial \phi} - \frac{\upsilon_\theta^2 + \upsilon_\phi^2}{r} \tag{2.15}$$

$$\alpha_\theta = \frac{\partial \upsilon_\theta}{\partial t} + \upsilon_r \frac{\partial \upsilon_\theta}{\partial r} + \frac{\upsilon_\theta}{r}\frac{\partial \upsilon_\theta}{\partial \theta} + \frac{\upsilon_\theta}{r\sin\theta}\frac{\partial \upsilon_\theta}{\partial \phi} + \frac{\upsilon_r \upsilon_\theta}{r} - \frac{\upsilon_\phi^2 \cot\theta}{r} \tag{2.16}$$

$$\alpha_\phi = \frac{\partial \upsilon_\phi}{\partial t} + \upsilon_r \frac{\partial \upsilon_\phi}{\partial r} + \frac{\upsilon_\theta}{r}\frac{\partial \upsilon_\phi}{\partial \theta} + \frac{\upsilon_\phi}{r\sin\theta}\frac{\partial \upsilon_\phi}{\partial \phi} + \frac{\upsilon_\phi \upsilon_r}{r} + \frac{\upsilon_\theta \upsilon_\phi \cot\theta}{r} \tag{2.17}$$

2.4 NAVIER–STOKES EQUATION

For Newtonian fluids, the viscous stress is proportional to the rate of shearing strain. Fig. 2.2 indicates the stresses in the x direction on a micro-cube of fluids. These stresses in Eqs. (2.9)–(2.11) may be expressed in terms of velocity gradient and fluid properties.

$$\tau_{xy} = \tau_{yx} = \mu\left(\frac{\partial \upsilon}{\partial x} + \frac{\partial u}{\partial y}\right)$$

$$\tau_{yz} = \tau_{zy} = \mu\left(\frac{\partial w}{\partial y} + \frac{\partial \upsilon}{\partial z}\right)$$

$$\tau_{zx} = \tau_{xz} = \mu\left(\frac{\partial u}{\partial z} + \frac{\partial w}{\partial x}\right)$$

$$\tau_{xx} = -\frac{2}{3}\mu(\nabla \cdot V) + 2\mu\frac{\partial u}{\partial x}$$

$$\tau_{yy} = -\frac{2}{3}\mu(\nabla \cdot V) + 2\mu\frac{\partial \upsilon}{\partial y}$$

$$\tau_{xx} = -\frac{2}{3}\mu(\nabla \cdot V) + 2\mu\frac{\partial w}{\partial z}$$

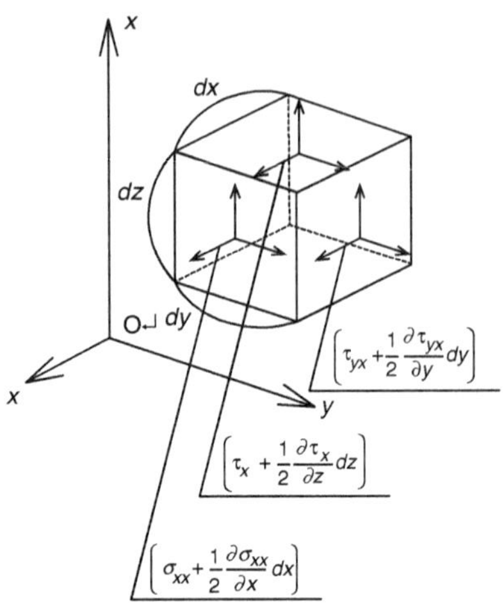

Figure 2.2 ***Stresses in the x direction on an element of fluids.***

where μ is viscosity of fluids. If fluids are incompressible fluid, the continuity equation is obtained as $(\nabla \cdot V) = 0$. By inserting these terms in Eqs. (2.9)–(2.11), the general equations of motion for incompressible Newtonian fluid in rectangular coordinates are obtained as follows,

$$\rho\left(\frac{\partial u}{\partial t} + u\frac{\partial u}{\partial x} + v\frac{\partial u}{\partial y} + w\frac{\partial u}{\partial z}\right) = \rho g_x - \frac{\partial p}{\partial x} + \mu\left(\frac{\partial^2 u}{\partial x^2} + \frac{\partial^2 u}{\partial y^2} + \frac{\partial^2 u}{\partial z^2}\right) \quad (2.18)$$

$$\rho\left(\frac{\partial v}{\partial t} + u\frac{\partial v}{\partial x} + v\frac{\partial v}{\partial y} + w\frac{\partial v}{\partial z}\right) = \rho g_y - \frac{\partial p}{\partial y} + \mu\left(\frac{\partial^2 v}{\partial x^2} + \frac{\partial^2 v}{\partial y^2} + \frac{\partial^2 v}{\partial z^2}\right) \quad (2.19)$$

$$\rho\left(\frac{\partial w}{\partial t} + u\frac{\partial w}{\partial x} + v\frac{\partial w}{\partial y} + w\frac{\partial w}{\partial z}\right) = \rho g_z - \frac{\partial p}{\partial z} + \mu\left(\frac{\partial^2 w}{\partial x^2} + \frac{\partial^2 w}{\partial y^2} + \frac{\partial^2 w}{\partial z^2}\right) \quad (2.20)$$

These equations of motion are called the Navier–Stokes equations for incompressible fluid with constant viscosity.

We get these equation expressed by vector symbolism,

$$\rho\frac{DV}{Dt} = \rho g - \nabla p + \mu\nabla^2 V \quad (2.21)$$

where ∇^2 is a Laplace operator and it is expressed as follows,

$$\nabla^2 = (\nabla \cdot \nabla) = \left(\frac{\partial^2}{\partial x^2} + \frac{\partial^2}{\partial y^2} + \frac{\partial^2}{\partial z^2}\right)$$

If we insert the vector formula, $\nabla\times[\nabla\times V]=\nabla(\nabla\cdot V)-\nabla^2 V$ in Eq. (2.21), the Navier–Stokes equations are rewritten as follows,

$$\rho\frac{DV}{Dt}=\rho g-\nabla p+\mu\{\nabla(\nabla\cdot V)-\nabla\times[\nabla\times V]\}$$

As the vorticity, $\Omega=[\nabla\times V]$ equals to zero for inviscid fluid flows, we get

$$\rho\frac{DV}{Dt}=\rho g-\nabla p \tag{2.22}$$

As a matter of course, Eq. (2.22) is the Euler's equation. Thus, an irrotational flow of incompressible fluids means a flow of ideal fluid flow.

The expressions of Eqs. (2.18)–(2.20) in each coordinates are given as follows,

Cylindrical coordinates:

$$\rho\left(\frac{\partial v_r}{\partial t}+v_r\frac{\partial v_r}{\partial r}+\frac{v_\theta}{r}\frac{\partial v_r}{\partial \theta}-\frac{v_\theta^2}{r}+v_z\frac{\partial v_r}{\partial z}\right)=\rho g_r-\frac{\partial p}{\partial r}$$
$$+\mu\left(\frac{\partial^2 v_r}{\partial r^2}+\frac{1}{r}\frac{\partial v_r}{\partial r}-\frac{v_r}{r^2}-\frac{2}{r^2}\frac{\partial v_\theta}{\partial \theta}+\frac{1}{r^2}\frac{\partial^2 v_r}{\partial \theta^2}+\frac{\partial^2 v_r}{\partial z^2}\right) \tag{2.23}$$

$$\rho\left(\frac{\partial v_\theta}{\partial t}+v_r\frac{\partial v_\theta}{\partial r}+\frac{v_\theta}{r}\frac{\partial v_\theta}{\partial \theta}+\frac{v_r v_\theta}{r}+v_z\frac{\partial v_\theta}{\partial z}\right)=\rho g_\theta-\frac{1}{r}\frac{\partial p}{\partial \theta}$$
$$+\mu\left(\frac{\partial^2 v_\theta}{\partial r^2}+\frac{1}{r}\frac{\partial v_\theta}{\partial r}-\frac{v_\theta}{r^2}+\frac{2}{r^2}\frac{\partial v_r}{\partial \theta}+\frac{1}{r^2}\frac{\partial^2 v_\theta}{\partial \theta^2}+\frac{\partial^2 v_\theta}{\partial z^2}\right) \tag{2.24}$$

$$\rho\left(\frac{\partial v_z}{\partial t}+v_r\frac{\partial v_z}{\partial r}+\frac{v_\theta}{r}\frac{\partial v_z}{\partial \theta}+v_z\frac{\partial v_z}{\partial z}\right)=\rho g_z-\frac{\partial p}{\partial z}$$
$$+\mu\left[\frac{1}{r}\frac{\partial}{\partial r}\left(r\frac{\partial v_z}{\partial r}\right)+\frac{1}{r^2}\frac{\partial^2 v_z}{\partial \theta^2}+\frac{\partial^2 v_z}{\partial z^2}\right] \tag{2.25}$$

Spherical coordinates:

$$\rho\left(\frac{\partial v_r}{\partial t}+v_r\frac{\partial v_r}{\partial r}+\frac{v_\theta}{r}\frac{\partial v_r}{\partial \theta}+\frac{v_\phi}{r\sin\theta}\frac{\partial v_r}{\partial \phi}-\frac{v_\theta^2+v_\phi^2}{r}\right)=\rho g_r-\frac{\partial p}{\partial r}$$
$$+\mu\left[\begin{array}{l}\dfrac{1}{r^2}\dfrac{\partial}{\partial r}\left(r^2\dfrac{\partial v_r}{\partial r}\right)+\dfrac{1}{r^2\sin\theta}\dfrac{\partial}{\partial \theta}\left(\sin\theta\dfrac{\partial v_r}{\partial \theta}\right)+\\ \dfrac{1}{r^2\sin^2\theta}\dfrac{\partial^2 v_r}{\partial \phi^2}-\dfrac{2v_r}{r^2}-\dfrac{2}{r^2}\dfrac{\partial v_\theta}{\partial \theta}-\dfrac{2v_\theta\cot\theta}{r^2}-\dfrac{2}{r^2\sin\theta}\dfrac{\partial v_\phi}{\partial \phi}\end{array}\right] \tag{2.26}$$

$$\rho\left(\frac{\partial \upsilon_r}{\partial t}+\upsilon_r\frac{\partial \upsilon_r}{\partial r}+\frac{\upsilon_\theta}{r}\frac{\partial \upsilon_r}{\partial \theta}+\frac{\upsilon_\varphi}{r\sin\theta}\frac{\partial \upsilon_r}{\partial \varphi}-\frac{\upsilon_\theta^2+\upsilon_\varphi^2}{r}\right)=\rho g_r-\frac{\partial p}{\partial r}$$

$$+\rho\left(\frac{\partial \upsilon_\theta}{\partial t}+\upsilon_r\frac{\partial \upsilon_\theta}{\partial r}+\frac{\upsilon_\theta}{r}\frac{\partial \upsilon_\theta}{\partial \theta}+\frac{\upsilon_\theta}{r\sin\theta}\frac{\partial \upsilon_\theta}{\partial \phi}+\frac{\upsilon_r\upsilon_\theta}{r}-\frac{\upsilon_\phi^2\cot\theta}{r}\right)$$
$$=\rho g_\theta-\frac{1}{r}\frac{\partial p}{\partial \theta}$$
$$+\mu\left[\begin{array}{l}\frac{1}{r^2}\frac{\partial}{\partial r}\left(r^2\frac{\partial \upsilon_\theta}{\partial r}\right)+\frac{1}{r^2\sin\theta}\frac{\partial}{\partial \theta}\left(\sin\theta\frac{\partial \upsilon_\theta}{\partial \theta}\right)+\\ \frac{1}{r^2\sin^2\theta}\frac{\partial^2 \upsilon_\theta}{\partial \phi^2}+\frac{2}{r^2}\frac{\partial \upsilon_r}{\partial \theta}-\frac{\upsilon_\theta}{r^2\sin^2\theta}-\frac{2}{r^2}\frac{\cos\theta}{\sin^2\theta}\frac{\partial \upsilon_\phi}{\partial \phi}\end{array}\right] \tag{2.27}$$

$$\rho\left(\frac{\partial \upsilon_\phi}{\partial t}+\upsilon_r\frac{\partial \upsilon_\phi}{\partial r}+\frac{\upsilon_\theta}{r}\frac{\partial \upsilon_\phi}{\partial \theta}+\frac{\upsilon_\phi}{r\sin\theta}\frac{\partial \upsilon_\phi}{\partial \phi}+\frac{\upsilon_\phi\upsilon_r}{r}+\frac{\upsilon_\theta\upsilon_\phi\cot\theta}{r}\right)$$
$$=\rho g_\phi-\frac{1}{r\sin\theta}\frac{\partial p}{\partial \phi}+$$
$$\mu\left[\begin{array}{l}\frac{1}{r^2}\frac{\partial}{\partial r}\left(r^2\frac{\partial \upsilon_\phi}{\partial r}\right)+\frac{1}{r^2\sin\theta}\frac{\partial}{\partial \theta}\left(\sin\theta\frac{\partial \upsilon_\phi}{\partial \theta}\right)+\\ \frac{1}{r^2\sin^2\theta}\frac{\partial^2 \upsilon_\phi}{\partial \phi^2}-\frac{\upsilon_\phi}{r^2\sin^2\theta}+\frac{2}{r^2\sin\theta}\frac{\partial \upsilon_r}{\partial \phi}+\frac{2}{r^2}\frac{\cos\theta}{\sin^2\theta}\frac{\partial \upsilon_\theta}{\partial \phi}\end{array}\right] \tag{2.28}$$

2.5 BOUNDARY CONDITIONS

It is required to set up the boundary conditions to obtain the analytical result of velocity profile by using the continuity equation and the equation of motion. In general, we can get two boundary conditions about the contact wall surface as follows;

1. The velocity component of vertical direction to wall surface is zero,
2. Fluid slip does not occur at wall surface.

For that of Euler's equation, (1) and the fluid slip conditions are used. However, fluid slip appears to be important in some non-Newtonian fluid

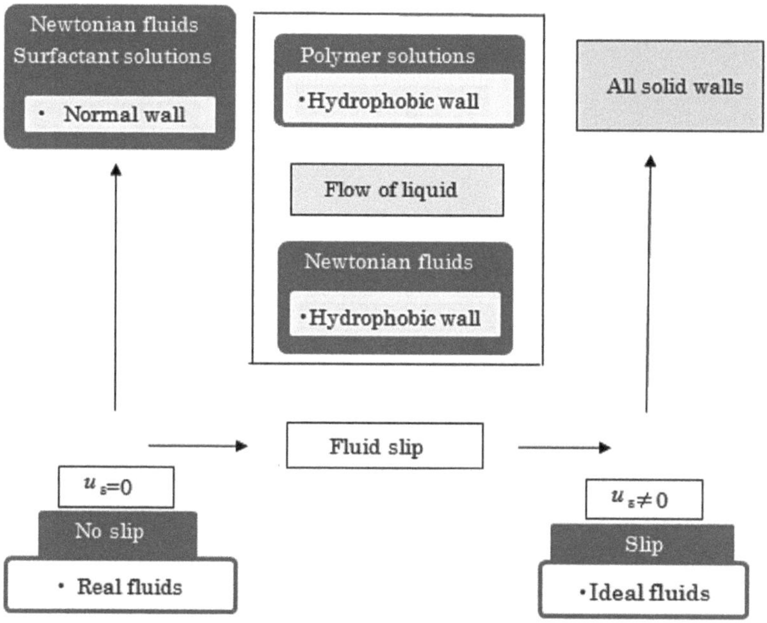

Figure 2.3 ***Boundary conditions at wall.***

flow or Newtonian fluid flow at some kinds of a highly water repellent wall as shown in Fig. 2.3.

Navier[1] gives the slip boundary conditions as following equation,

$$u_s = \frac{\mu}{\beta}\left|\frac{\partial u}{\partial y}\right| = \frac{\tau_w}{\beta} \qquad (2.29)$$

where u and u_s are the velocity and the slip velocity, and β and μ are sliding constant and viscosity, respectively. c and τ_w are the distance from the wall and the shear stress. The sliding constant in Eq. (2.29) is determined from the relationship between u_s and τ_w if we can measure these values by experiment.

Fig. 2.4 shows an example of the experimental results for Newtonian fluids[2] and polymer solutions[3] in a circular pipe with a highly water repellent wall. The sliding constant β is the constant value for Newtonian fluid, in other word, the slip velocity is proportional to the shear stress, but it is nonlinear for polymer solutions.

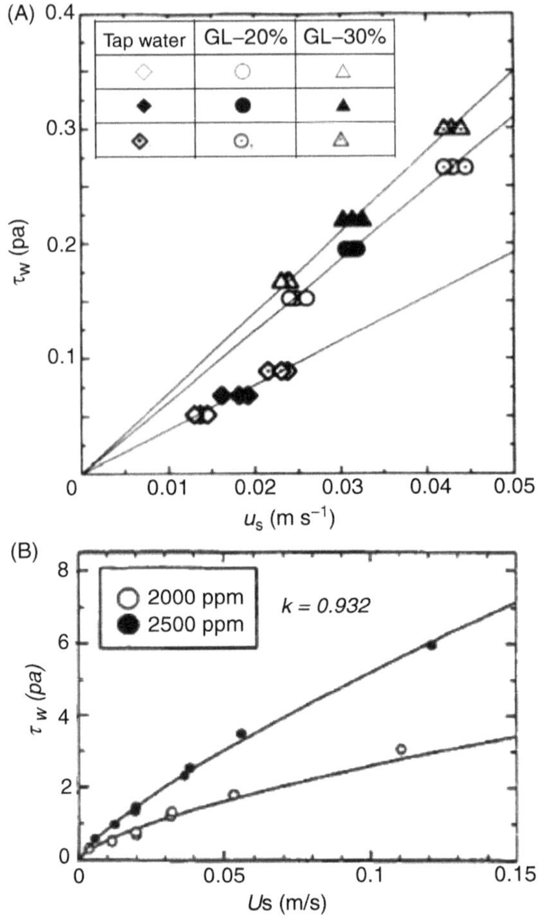

Figure 2.4 ***Slip velocity at a highly water-repellent wall.*** (A) Newtonian fluids. (B) Polymer solutions.

Example 2.1 Steady laminar flow in a rectangular duct

As shown in Fig. 2.5, incompressible fluids of density, ρ and viscosity, μ flow in a straight of uniform duct section with the width, w and the height, H under a constant pressure condition, p. It is a fully developed steady flow. Determine the velocity profile of laminar flow in the duct.

Solution

For this flow system all terms of the equation of continuity as written rectangular coordinates are zero, and x and y components of the equation of motion Eqs. (2.18) and (2.19) reduce to

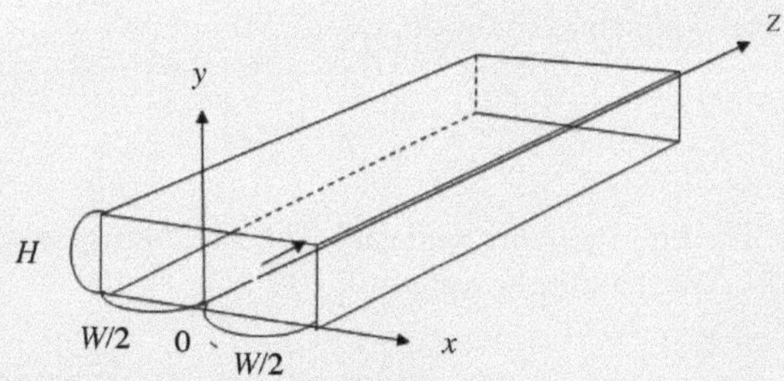

Figure 2.5 ***Rectangular coordinates for a duct.***

$$\frac{\partial p}{\partial x} = \frac{\partial p}{\partial y} = 0 \tag{2.30}$$

Eq. (2.20) becomes

$$\mu\left(\frac{\partial^2 w}{\partial x^2} + \frac{\partial^2 w}{\partial y^2}\right) = \frac{dp}{dz} \tag{2.31}$$

If fluid slip does not occur at the wall, the boundary condition reduce to

$$w(x,H) = w(x,0) = w\left(\pm\frac{W}{2}, y\right) = 0 \tag{2.32}$$

To solve Eq. (2.31), the velocity, w is assumed to

$$w = \phi(x,y) + V(y) \tag{2.33}$$

By substituting Eq. (2.33) in Eq. (2.31), the following equation may be obtained,

$$\left(\frac{\partial^2 \phi}{\partial x^2} + \frac{\partial^2 \phi}{\partial y^2} + \frac{\partial^2 V}{\partial y^2}\right) = \frac{1}{\mu}\frac{dp}{dz} \tag{2.34}$$

On the other hand, we assume V in which satisfies the following equation,

$$\frac{d^2 V}{dy^2} = \frac{1}{\mu}\left(\frac{dp}{dz}\right) \tag{2.35}$$

Thus, Eq. (2.34) becomes

$$\nabla^2 \phi = \left(\frac{\partial^2 \phi}{\partial x^2} + \frac{\partial^2 \phi}{\partial y^2}\right) = 0 \tag{2.36}$$

By using the boundary condition, Eq. (2.33) reduces to

$$V = \frac{y}{2\gamma}\left(-\frac{dp}{dz}\right)(H - y) \tag{2.37}$$

We can solve Eq. (2.36) by separation of variables. If we assume the form of $\phi = X(x)Y(y)$ and substitute in Eq. (2.36),

$$\frac{X''}{X} = -\frac{Y''}{Y} = \lambda$$

where λ must be a constant value. Thus, the general solutions for X and Y will be given as

$$X = Ae^{\sqrt{\lambda}x} + Be^{-\sqrt{\lambda}x}$$
$$Y = C\cos\sqrt{\lambda}y + D\sin\sqrt{\lambda}y$$

The boundary conditions for ϕ are

$$w(x,0) = 0; \quad \phi(x,0) = 0$$
$$w(x,H) = 0; \quad \phi(x,0) = 0$$
$$w\left(\pm\frac{W}{2}, y\right) = 0; \quad \phi\left(\pm\frac{W}{2}, y\right) = \phi(y) = -\frac{y}{2\mu}\left(\frac{dp}{dz}\right)(H - y)$$

If we take account for $Y(0) = 0$, $Y(H) = 0$ and $X\left(\pm\frac{W}{2}\right) = \phi(y)$, $C = 0$ and $\sin\sqrt{\lambda}H = 0$ are obtained and then, $\sqrt{\lambda} = (n\pi)/H$ is reduced. ϕ is given as

$$\phi = \sum_{n=1}^{\infty} A_n \sin\left(\frac{n\pi}{H}y\right)\cosh\left(\frac{n\pi W}{2H}\right) \tag{2.38}$$

If we replace $A_n = A_n \cosh\left(\frac{n\pi W}{2H}\right)$, it becomes

$$\phi = \sum_{n=1}^{\infty} A_n \sin\left(\frac{n\pi}{H}y\right)$$

We obtained the following equation by using Fourier transform

$$A_n = \frac{2}{H}\int_0^H \phi \sin\frac{n\pi}{H}y\,dy$$

By substituting $\phi = -\frac{y}{2\mu}\left(-\frac{dp}{dz}\right)(H - y)$, $A_n = 4\left(\frac{H}{n\pi}\right)^3$ $(n = 1, 3, 5, \cdots\cdots)$ is reduced.

We have

$$\phi = \frac{4H^2}{\pi^3\mu}\left(\frac{dp}{dz}\right)\sum_{n=1,3,5,\cdots}^{\infty}\frac{1}{n^3}\sin\left(\frac{n\pi}{H}y\right)\frac{\cosh\left(\frac{n\pi}{H}x\right)}{\cosh\left(\frac{n\pi W}{2H}\right)} \tag{2.39}$$

Consequently, the velocity profile in a rectangular duct is given

$$w = \frac{y}{2\mu}\left(-\frac{dp}{dz}\right)(H-y) - \frac{4H^2}{\pi^3\mu}\left(-\frac{dp}{dz}\right)\sum_{n=1,3,5,\cdots}^{\infty}\frac{1}{n^3}\sin\left(\frac{n\pi}{H}y\right)\frac{\cosh\left(\frac{n\pi}{H}x\right)}{\cosh\left(\frac{n\pi W}{2H}\right)} \tag{2.40}$$

Example 2.2 Momentum integral equation of flow around a rotating disk

The momentum integral equation of flow around a rotating disk which is shown in Fig. 2.6, is derived as following equations using the momentum and angular momentum conservation law to the boundary layer that r and θ-component of the velocity are v_r and v_θ. And then, τ and δ are the shear stress and the boundary layer thickness, respectively.

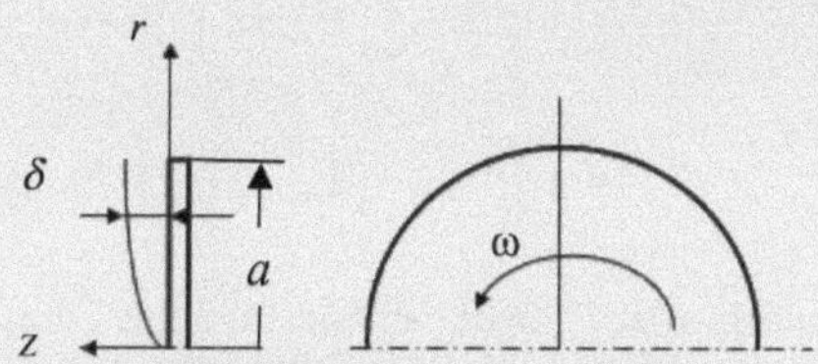

Figure 2.6 ***Flow around a rotating disk.***

Show that these equations is given also for Newtonian fluids by integrating Navier–Stokes equation for the interval 0 to δ

$$\frac{d}{dr}\left(r\int_0^\delta v_r^2\,dz\right) - \int_0^\delta v_\theta^2\,dz = -\frac{r}{\rho}\tau_{rz|z=0} \tag{2.41}$$

$$\frac{d}{dr}\left(r^2\int_0^\delta v_r v_\theta\,dz\right) = -\frac{r^2}{\rho}\tau_{\theta z|z=0} \tag{2.42}$$

Solution

Navier–Stokes equations and the continuity equations are set up the following equations as it is an axial symmetry flow

$$\rho\left(\upsilon_r\frac{\partial \upsilon_r}{\partial r}-\frac{\upsilon_\theta^2}{r}+\upsilon_z\frac{\partial \upsilon_r}{\partial z}\right)=\mu\frac{\partial^2 \upsilon_r}{\partial z^2} \tag{2.43}$$

$$\rho\left(\upsilon_r\frac{\partial \upsilon_\theta}{\partial r}+\frac{\upsilon_r\upsilon_\theta}{r}+\upsilon_z\frac{\partial \upsilon_\theta}{\partial z}\right)=\mu\frac{\partial^2 \upsilon_\theta}{\partial z^2} \tag{2.44}$$

$$\frac{1}{r}\frac{\partial}{\partial r}\left(r\upsilon_r\right)+\frac{\partial \upsilon_z}{\partial z}=0 \tag{2.45}$$

where $\upsilon_r, \upsilon_\theta, \upsilon_z$ are the r, θ, z components of velocity in the boundary layer, respectively. It is assumed that they satisfy the no-slip boundary conditions and υ_r and υ_θ varnish at the outer of the boundary layer.

For the r-component of the integral momentum equation, we integrate Eq. (2.43) for the interval 0 to δ after the multiplication of r to the both sides of the equation.

$$\begin{aligned}
&\rho\int_0^\delta r\left(\upsilon_r\frac{\partial \upsilon_r}{\partial r}-\frac{\upsilon_\theta^{\,2}}{r}+\upsilon_z\frac{\partial \upsilon_r}{\partial z}\right)dz\\
&=\rho\left[\int_0^\delta\left(\frac{r}{2}\frac{\partial \upsilon_r^{\,2}}{\partial r}-\upsilon_\theta^{\,2}\right)dz+r\upsilon_r\upsilon_z\Big|_0^\delta-\int_0^\delta r\upsilon_r\left(\frac{\partial \upsilon_z}{\partial z}\right)dz\right]\\
&=\rho\left[\int_0^\delta\left(\frac{r}{2}\frac{\partial \upsilon_r^{\,2}}{\partial r}-\upsilon_\theta^{\,2}\right)dz+\int_0^\delta \upsilon_r^{\,2}\,dz+\int_0^\delta\frac{r}{2}\frac{\partial \upsilon_r^{\,2}}{\partial r}dz\right]\\
&\rho\left[\int_0^\delta\frac{\partial}{\partial r}\left(r\upsilon_r^{\,2}\right)dz-\int_0^\delta \upsilon_\theta^{\,2}\,dz\right]
\end{aligned} \tag{2.46}$$

By using the boundary layer approximation that $\tau_{rz}\big|_{z=\delta}=0$ and $\upsilon_r\big|_{z=\delta}=0$ are satisfied and δ depends on r only, we can permute of the calculation of the derivation and the integration from the Leibniz formula.

Thus, the r component of the momentum integral equation, which is given by Eq. (2.41), is derived as follows,

$$\frac{d}{dr}\left(r\int_0^\delta \upsilon_r^2dz\right)-\int_0^\delta \upsilon_\theta^2\,dz=\frac{1}{\rho}\int_0^\delta r\mu\left(\frac{\partial^2 \upsilon_r}{\partial z^2}\right)dz=\frac{r}{\rho}\left|\mu\left(\frac{\partial \upsilon_r}{\partial z}\right)\right|_{z=0}^{z=\delta}=-\frac{r}{\rho}\tau_{rz}\big|_{z=0}$$

For the θ-component of the integral momentum equation, we integrate for the interval 0 to δ after multiplying Eq. (2.44) by r^2 to the both side of the equation.

$$\rho\int_0^\delta\left(r^2v_r\frac{\partial v_\theta}{\partial r}+rv_rv_\theta+r^2v_z\frac{\partial v_\theta}{\partial z}\right)dz$$
$$=\rho\int_0^\delta\left(r^2v_r\frac{\partial v_\theta}{\partial r}+rv_rv_\theta\right)dz-\rho\left[r^2v_\theta v_r\Big|_{z=0}^{z=\delta}-r^2\int_0^\delta v_\theta\left(\frac{\partial v_z}{\partial z}\right)dz\right] \quad (2.47)$$

Substituting Eq. (2.45) into Eq. (2.47) yields

$$\rho\int_0^\delta\left[r^2v_r\frac{\partial v_\theta}{\partial r}+rv_rv_\theta+rv_\theta\frac{\partial}{\partial r}(rv_r)\right]dz=\rho\int_0^\delta\frac{\partial}{\partial r}\left(r^2v_rv_\theta\right)dz$$

From the boundary layer approximation, $v_{r|z=\delta}=v_{\theta|z=\delta}=0$ and $\tau_{\theta z|z=\delta}=0$, we obtain

$$\frac{d}{dr}\left[r^2\int_0^\delta\left(v_rv_\theta\right)dz\right]=\frac{r^2}{\rho}\int_0^\delta\mu\left(\frac{\partial^2v_\theta}{\partial z^2}\right)dz=\frac{r^2}{\rho}\left|\mu\frac{\partial v_\theta}{\partial z}\right|_{z-0}^{z-\delta}=-\frac{r^2}{\rho}\tau_{\theta z}\Big|_{z=0}$$

As expected, we see that these equations are identical.

Although these equations are derived using Navier–Stokes equation, we can obtain the same results for turbulent boundary layer using the time smoothed velocity, pressure and stress in the motion of equations. Thus, the frictional drag of a rotating disk will be deduced from these equations by assuming the velocity profiles of the boundary layer. Tomita and Mochimaru[4] deduced the equation for viscoelastic fluids.

2.6 POWER-LAW FLUIDS

We must give the rheological equation to reduce the equation of motion for non-Newtonian fluids. As described above, Newtonian fluids play a linear behavior in the relationship between the stresses and shear rate. However, that of time independent complex mixtures is a nonlinear as shown in Fig. 2.7. If we rewrite these flow curves using logarithmically scale, they show straight lines. Thus, these lines may be predicted by the following equation

$$\tau=\tau_0+K\left(\dot{\gamma}\right)^n. \quad (2.48)$$

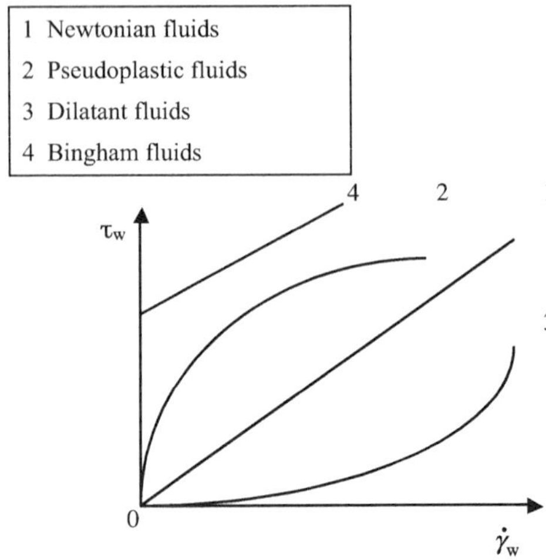

Figure 2.7 ***Flow curve of time independent fluids.***

If the yield stress, τ_0 is zero

$$\tau = K\left(\dot{\gamma}\right)^{n} = K\left(\dot{\gamma}\right)^{n-1}\left(\dot{\gamma}\right) = \eta_{ck}\dot{\gamma} \qquad (2.49)$$

It is so-called power law fluids, in this equation K is the consistency index or power law coefficient, and n is the flow behavior index or power law exponent. η_{ck} is the apparent viscosity.

Eq. (2.49) is used widely in an engineering calculation because the approximation is a simple. The flow behavior index, n, is determined as the gradient of a plot of τ_w versus $\dot{\gamma}$ on logarithmic coordinates. In the case of the value of n is less than unity, it is called pseudo-plastic fluids. This type of behavior is widely encountered in polymer solutions or fine solid particle suspensions. Although this equation is simple any questions remain in the application. As is clear from Eq. (2.49), it fails both at a very low and very high shear rate. When the shear rate approaches zero, the consistency index approaches an infinite value, and it approaches zero when the shear rate approaches an infinite value. Actual experimental data conform to the equation for Newtonian fluids at very low shear rate, and data become constant. And then, the dimension of the viscosity

Example 2.3 Determination of physical constant of power law fluids

Show the flow curve creation method obtained by use of a capillary viscometer.

Solution

Flow curve indicates a relationship between the shear stress, τ and the shear rate, $\dot{\gamma}$. The integral method that the rheological equation is presumed is widely used in general as it is a simple compare with the deferential method.

For incompressible steady and fully developed flow in a circular pipe with the radios, a, the Navier–Stokes equations, Eqs. (2.23)–(2.25) are described by cylindrical coordinates become

$$\left(\frac{\partial p}{\partial r}\right)=\left(\frac{\partial p}{\partial \theta}\right)=0 \tag{2.50}$$

$$0=-\left(\frac{dp}{dz}\right)+\mu\frac{1}{r}\frac{\partial}{\partial r}\left(r\frac{\partial \upsilon_z}{\partial r}\right) \tag{2.51}$$

By solving Eq. (2.51) under the boundary conditions without fluid slip

$$\upsilon_z=\frac{a^2}{4\mu}\left(-\frac{dp}{dz}\right)\left[1-\frac{r^2}{a^2}\right] \tag{2.52}$$

If the pressure deference of the pipe length l, is described Δp, the pressure gradient is given

$$\left(-\frac{dp}{dz}\right)=\frac{\Delta p}{l}$$

The flow rate is calculated as

$$Q=\int_0^{2\pi}\int_0^{a} r\upsilon_z\,dr\,d\theta=\frac{\pi a^4\Delta p}{8\mu l}$$

The shear stress τ_w and the shear rate $\dot{\gamma}_w$ at the wall, are calculated

$$\tau_w=\mu\left(-\frac{d\upsilon_z}{dr}\right)_{r=a}=\frac{a}{2}\left(\frac{\Delta p}{l}\right) \tag{2.53}$$

$$\dot{\gamma}_w=\left(-\frac{d\upsilon_z}{dr}\right)_{r=a}=\frac{a}{2\mu}\left(\frac{\Delta p}{l}\right)=\left(\frac{8U_m}{d}\right) \tag{2.54}$$

where U_m is the mean velocity given as $U_m=\dfrac{Q}{\pi a^2}=\dfrac{a^2\Delta p}{8\mu l}$

Thus, we can plots of τ_w and $\dot{\gamma}_w$ by measuring the flow rate and the pressure deference of the capillary.

However, we must pay attention that the power law exponent obtained experimentally by this method does not equal exactly with that of the rheological equation of power law fluids.

The equation of motion for power law fluids becomes

$$0 = -\frac{dp}{dz} - \frac{1}{r}\left(\frac{d}{dr}(r\tau_{rz})\right) \tag{2.55}$$

From Eq. (2.49), the rheology equation of power law fluids is given as

$$\tau_{rz} = K\left(-\frac{dv_z}{dr}\right)^n \tag{2.56}$$

We substitute Eq. (2.56) into Eq. (2.55), and the velocity profile is deduced,

$$v_z = \frac{n}{(n+1)}\left[\frac{1}{2K}\left(-\frac{dp}{dz}\right)\right]^{1/n} a^{(n+1)/n}\left[1-\left(\frac{r}{a}\right)^{(n+1)/n}\right] \tag{2.57}$$

The flow rate, Q is

$$Q = \int_0^{2\pi}\int_0^a r v_z \, dr \, d\theta = \frac{2\pi n}{(n+1)}\left[\frac{1}{2K}\left(-\frac{dp}{dz}\right)\right]^{1/n} a^{(n+1)/n}\int_0^a r\left[1-\left(\frac{r}{a}\right)^{(n+1)/n}\right]dr$$
$$= \frac{2\pi a^2 n}{(n+1)}\left[\frac{1}{2K}\left(-\frac{dp}{dz}\right)\right]^{1/n} a^{(n+1)/n}\left[\frac{n+1}{2(3n+1)}\right] \tag{2.58}$$

The mean velocity, U_m

$$U_m = \left(\frac{Q}{\pi a^2}\right) = \frac{n}{(3n+1)}\left[\frac{1}{2K}\left(-\frac{dp}{dz}\right)\right]^{1/n} a^{(n+1)/n} \tag{2.59}$$

The shear rate of power law fluids at the wall is reduced as

$$\dot{\gamma}_w = \frac{(3n+1)}{4n}\left(\frac{8U_m}{d}\right) \tag{2.60}$$

Thus, if it is power law fluids, we must calculate the shear rate by applying Eq. (2.60) instead $\dot{\gamma}_w = \left(\frac{8U_m}{d}\right)$. However, because n is determined by the gradient of the flow curve having the logarithmic scale and it is unknown at that stage, we cannot obtain n by using Eq. (2.60). Eq. (2.60) may be used for the modification. Thus, the power law exponent determined experimentally from the flow curve may be distinguished by writing it as n' according to the strict definition.

Fig. 2.8 shows an example of a capillary viscometer. The constant flow rate, Q is obtained using a piston with a constant speed controlled by a P.C. after a test fluid was filled in the syringe, and the pressure difference, Δp of a capillary is measured by a pressure transducer. The flow curves of xanthan gum 50–500 ppm solutions are shown in Fig. 2.9. These values of solutions were approximately determined from the solid lines in the figure. The reader must pay attention to the fact that the approximate value will be arranged in the range of shear rate.

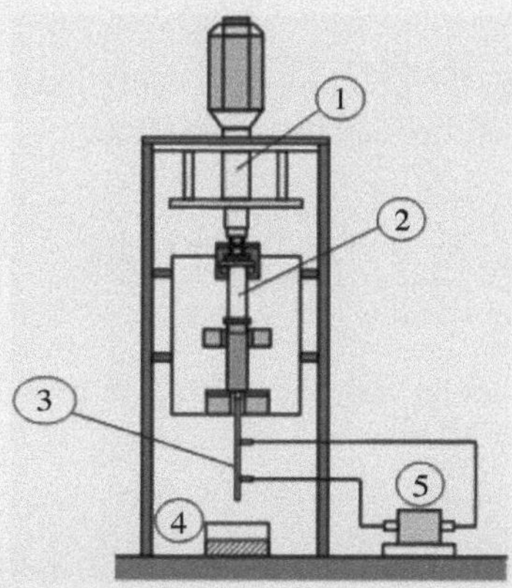

Figure 2.8 ***Capillary viscometer.*** (1) Piston, (2) syringe, (3) capillary, (4) tank, and (5) pressure transducer.

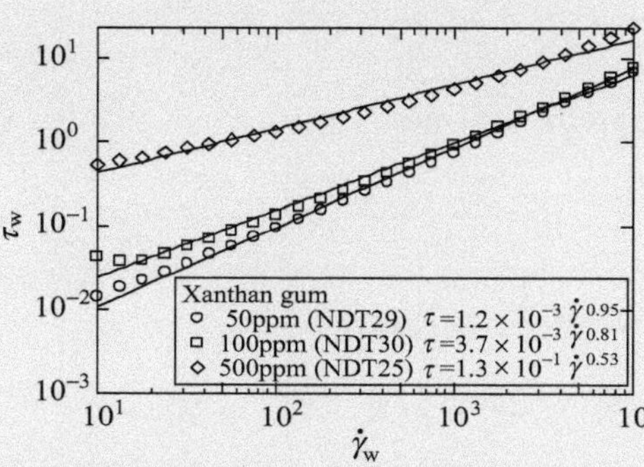

Figure 2.9 ***Flow curve of xanthan gum solutions.***

Example 2.4 Determination of a velocity profile with fluid slip
Obtain an expression for a velocity profile with fluid slip in a circular pipe using Navier's hypothesis, Eq. (2.29). Fluids are an incompressible fluid.

Solution
Navier-Stokes equation for cylindrical coordinates, given in Eqs. (2.23)–(2.25), were reduced by setting the flow conditions that it is fully developed flow

$$0 = \frac{\partial p}{\partial r} = \frac{\partial p}{\partial \theta} \tag{2.61}$$

$$0 = -\frac{\partial p}{\partial z} + \mu\left\{\frac{1}{r}\frac{\partial}{\partial r}\left(r\frac{\partial \upsilon_z}{\partial r}\right)\right\} \tag{2.62}$$

Eq. (2.61) gives that the pressure, p is $p=p(z)$, and the velocity profile υ_z is $\upsilon_z = \upsilon_z(r)$ because of the fully developed flow.

Thus,

$$\frac{1}{r}\frac{d}{dr}\left(r\frac{d\upsilon_z}{dr}\right) = \frac{1}{\mu}\left(\frac{dp}{dz}\right) \tag{2.63}$$

Eq. (2.63) may be integrated immediately to give

$$\upsilon_z = \frac{1}{4\mu}\left(\frac{dp}{dz}\right)r^2 + \mathrm{C}_1\,\mathrm{In}\,\mathrm{r} + \mathrm{C}_2$$

in which C_1 and C_2 are constants of integration. Because υ_z cannot be infinite at $r = 0$, the constant C_1 is set equal to zero. If the slip velocity at $\mathrm{r} = \mathrm{a}$ equals u_s, the constant C_2 is reduced.

$$\mathrm{C}_2 = \left\{u_s - \frac{1}{4\mu}\left(\frac{dp}{dz}\right)a^2\right\}$$

The velocity profile becomes

$$\upsilon_z = \frac{1}{4\mu}\left(-\frac{dp}{dz}\right)a^2\left(1 - \frac{r^2}{a^2}\right) + u_s \tag{2.64}$$

Navier' hypothesis, Eq. (2.29) gives

$$\tau_w = \mu\left(-\frac{d\upsilon_z}{dr}\right)_{r=a} = \beta \mathrm{u}_s$$

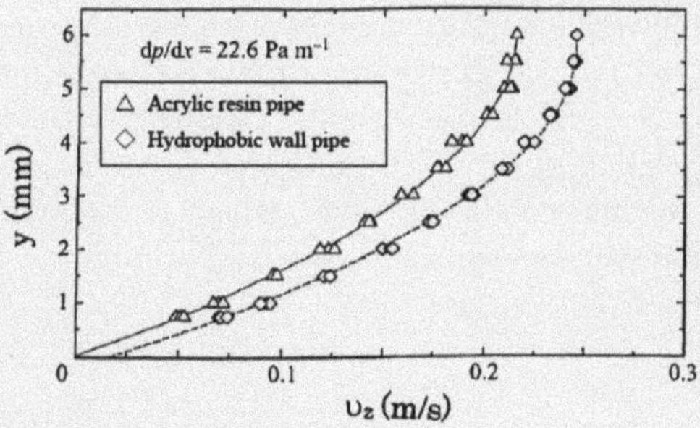

Figure 2.10 ***Velocity profile with slip velocity.***

We obtain the velocity profile with fluid slip

$$\upsilon_z = \frac{1}{4\mu}\left(-\frac{dp}{dz}\right)a^2\left(1-\frac{r^2}{a^2}\right)+\frac{a}{2\beta}\left(-\frac{dp}{dz}\right) \tag{2.65}$$

We know that the difficulty remains in the determination for the sliding constant, β, however the order is examined by resent study[5]. Fig. 2.10 shows the comparison between Eq. (2.65) and the experimental data.

When it is the no-slip condition, $\beta \to \infty$, and Eq. (2.65) becomes the Hagen-Poiseuillie flow

$$\upsilon_z = \frac{1}{4\mu}\left(-\frac{dp}{dz}\right)a^2\left(1-\frac{r^2}{a^2}\right)$$

This result reduces the friction factor.

In a fully developed flow, the pressure gradient $\left(-\frac{dp}{dz}\right)$ is constant. Therefore, the pressure gradient is $(-dp/dz) = (p_1 - p_2)/l = \Delta p/l$.

The volume flow rate Q can be obtained as follows:

$$Q = \int_0^a 2\pi r \upsilon_z \, dr = \frac{\pi a^4 \Delta p}{8\mu l}\left(1+\frac{4\mu}{a\beta}\right)$$

If a friction factor is defined as $\lambda = 4\left[\tau_w / \left(\rho U_0^{\,2}/2\right)\right]$, where ρ is fluid density, then in combination with Eq. (2.29), the friction factor λ becomes

$$\lambda = \frac{64}{\mathrm{Re}}\frac{1}{\left[1+\left(\frac{4\mu}{a\beta}\right)\right]} = \frac{64}{\mathrm{Re}}\frac{1}{(1+4S)} \tag{2.66}$$

Thus, in a fully developed laminar flow with fluid slip, the friction factor is a function of, not only the Reynolds number Re but also the nondimensional parameter $S = (\mu/\alpha\beta)$. Because the parameter related to fluid slip is always a positive value in Eq. (2.66), it decreases compared with that of a flow under the no-slip conditions $S = 0$, which is given as $\beta \to \infty$ in Eq. (2.29). In other words, laminar drag reduction occurs in a flow with fluid slip.

does not consist of an integer because the power law exponent is not an integer in most cases.

REFERENCES

1. Navier CMLH. *Mémoires de l'Académie Royale des Sciences de l'Institut de France* 1823;**6**:389.
2. Watanabe K, Yanuar, Udagawa H. *J. Fluid Mech.* 1999;**381**:225.
3. Watanabe K, Udagawa H. *AIChE J.* 2001;**26**:729.
4. Tomita Y, Mochimaru Y. *Bull JSME* 1973;291.
5. Watanabe K. *Laminar Drag Reduction.* Bentham Publishers Co. Ltd; 2015 p. 18.

CHAPTER 3

Drag Reduction in a Heterogeneous Flow

Contents

3.1 Introduction 35
3.2 Liquid–Solid Two-Phase Flow 36
3.3 Gas–Solid Two-Phase Flow 57
References 74

3.1 INTRODUCTION

According to the increase of the size of matter, which was added into fluids, the flow becomes a heterogeneous flow. For example, the liquid–solid mixtures will be a two-phase flow in pipes when the particle mean diameter is an order more than several tens of micron. Heterogeneous flow appears many a time in a pipeline system on hydraulic transport of coal or pneumatic conveying of grain. In recent year, the systems were grown in size and the technique has been applied in various industrial fields, because it is a continuous transport method and the environmental load is small.

The difference of flow behavior of liquid–solid and gas–solid two-phase flow in pipes, depends in good part on the physical parameter, the density ratio between fluids and solid particles. In general speaking, the particles flow while following fluids since the ratio is O (1) in the order. However, we cannot negligible the inertia force acting on the particles in fluids because that of gas–solid two-phase flow is O (10^{-3}) in the order. Thus, they are a complicate flow and there are still some things we must settle in the industrial applications.

The recent trends for a hydraulic transport system are a high concentration and a low speed transport to avoid wear of pipeline. In the treatment, the drag reduction technique in multiphase flows might be noticed from viewpoint of energy saving for the system driving power.

In this chapter, some experimental data of a horizontal pipe flow system related multiphase flows, are represented and discussed.

Drag Reduction of Complex Mixtures.
http://dx.doi.org/10.1016/B978-0-12-809920-9.00003-2

3.2 LIQUID–SOLID TWO-PHASE FLOW

Fig. 3.1 shows the pattern diagram for the typical data of pressure loss of solid–liquid two-phase flow in a horizontal circular pipe with the flow pattern diagrams. In the figure, the values of the vertical and the horizontal axis are the pressure loss per unit length, $\Delta p/l$ and the mean velocity of two-phase flow, υ_m, respectively. The solid line means the value of water obtained by Blasius formula. As described in Chapter 1, it is Δp_a with (+) in Eq. (3.1) for this flow field. According the increase of the concentration, the pressure loss increases, and also according to the increase of the mean velocity, the particles are dispersed in a pipe.

On the other hand, according to decrease of the mean velocity, the pressure loss decreases however the value increases inversely with decrease of the velocity. In other words, the minimum value appears in the pressure loss data. In this case, a sliding layer of the particle is formed at a bottom of the pipe as shown the flow patterns in Fig. 3.1. In the hydraulic transport pipeline systems, the minimum value is used as a target for the optimum transport velocity in general. Thus, the reduction of the pressure loss and velocity of the flow ranges with the sliding layer is a target for the flow system with high efficiency.

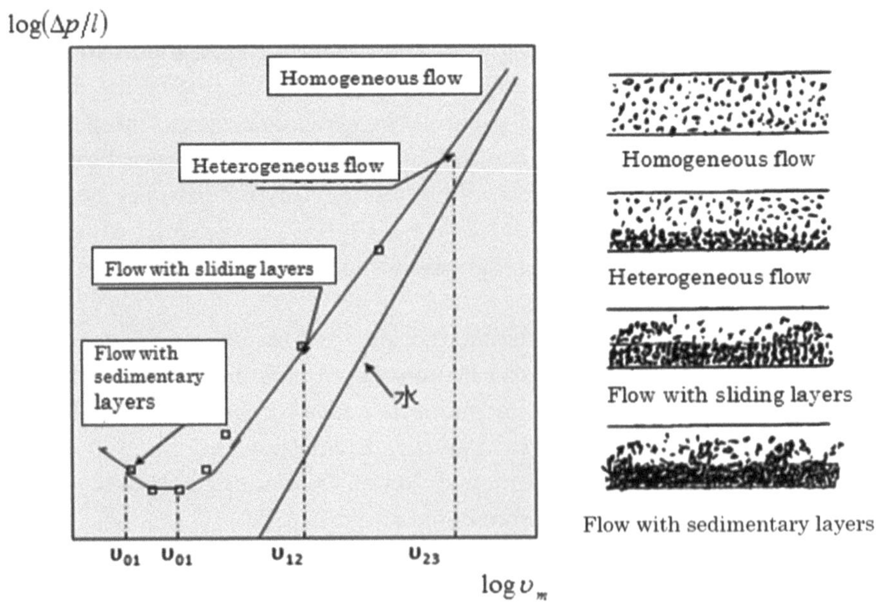

Figure 3.1 *Pressure loss and flow patterns of solid–liquid two-phase flow.*

For the case of a horizontal pipe flow, Eq. (3.1) in Chapter 1 will be rewritten as follows by considering a particle motion in a pipe,

$$\Delta p_t = \Delta p_f + \Delta p_a = \Delta p_f + (\Delta p_{a1} + \Delta p_{a2} + \Delta p_{a3} + \Delta p_{a4}) \tag{3.1}$$

where Δp_a is the additional pressure, and it is given as

$$\Delta p_a = (\Delta p_{a1} + \Delta p_{a2} + \Delta p_{a3} + \Delta p_{a4}) \tag{3.2}$$

Each term of the additional pressure $\Delta p_{a1} - \Delta p_{a4}$ will depend on the including particle behavior in a pipe. Such as, Δp_{a1} is the loss between particles and pipe wall surface, Δp_{a2} is the drag of particles based the velocity difference between fluid and the particle, Δp_{a3} is the friction loss by collision of particles to each other and Δp_{a4} is the loss by other reasons. As a matter of course, the order of each pressure loss varies according to the flow patterns

In the case of heterogeneous flow that it is millimeter order in particle size, since the physical constant of carrier fluid does not change by adding particle and the turbulence modification does not occur in most cases, the reduction of Δp_{a1} in Eq. (3.2) is a target in order to obtain the drag reduction phenomenon. Therefore, the drag reduction may be obtained if we can change the flow pattern in a pipe from the stationary bed of particles to the sliding layer.

One of the techniques is an application of deformed pipe with a spiral rib[1] or a spiral wall[2] for a transport pipe. Particles float in these pipes by a swirling flow, and they change from the stationary layer to sliding layer in the low velocity range of the complex mixtures. As the results, we can obtain the drag reduction phenomenon for that case. As the example which we can apply to liquid–solid[3,4] and gas–solid two-phase flows,[5] a spiral pipe can be mentioned.

As an example, Fig. 3.2A and B show the cross section and the appearance of a spiral pipe with the dimensions of Type III pipe, respectively. Although three types of spiral pipe that the ratio between the pith and the maximum diameter, (p/D_0) is 3.5, 7.0, and 12.0 were tested for liquid–solid two-phase flow, $(p/D_0) = 7.0$ was the most effective pipe to obtain the drag reduction.

The flow behavior in the spiral pipe are shown by the flow visualization and the measurement of velocity profile. Fig. 3.3 is an example of the streak of a flow obtained by dye injection method at R_e = 1230. In the figure, the variations of the streak in the direction normal the pipe wall are shown. The fluids between the wall and the nondimensional distance, d/D_i' = 0.2 flow along the pipe wall, but the streak at d/D_i' = 0.4 is parallel to the pipe axis.

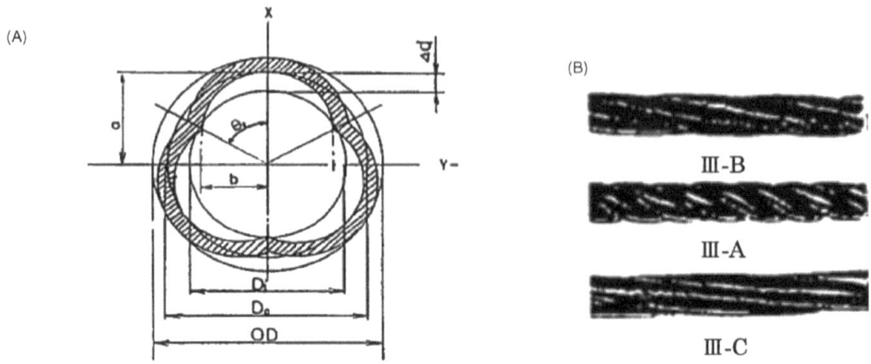

Pipe	D_0	D_i	D_e	d	p	p/D_0
Ⅲ-A	44.5	34.5	36.9	5.0	155	3.5
Ⅲ-B	44.4	33.8	37.0	5.3	310	7.0
Ⅲ-C	43.0	33.6	38.0	4.5	510	12.0

Figure 3.2 ***Outline of spiral pipes.*** (A) Cross section and (B) appearance.

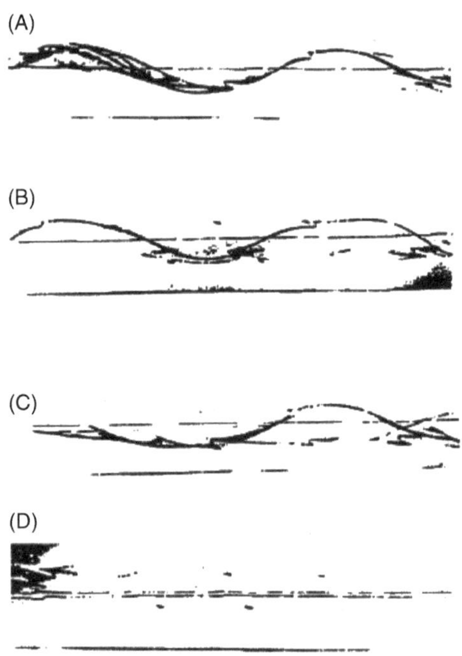

Figure 3.3 ***Flow streaks in spiral pipe.*** (A) $R_e = 1237$ $d/D_i' = 0.05$. (B) $R_e = 1232$ $d/D_i' = 0.1$. (C) $R_e = 1232$ $d/D_i' = 0.2$. (D) $R_e = 1232$ $d/D_i' = 0.4$.

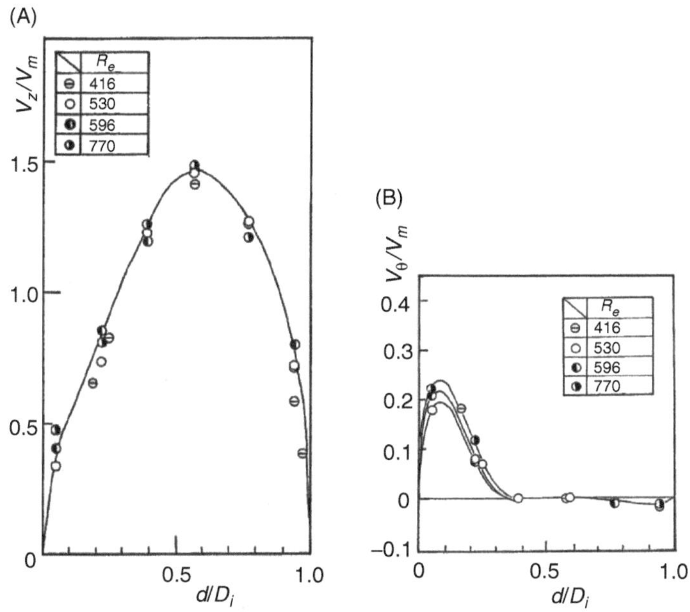

Figure 3.4 ***Velocity profiles in laminar region.*** (A) Axial velocity component. (B) Circumferential velocity component.

Thus the flow has no swirl component in the range from the pipe axis to $d/D_i' = 0.4$.

Other hand, the velocity profiles are measured by using a three holes type Pitot probe which is to measure the three dimensional flow. Fig. 3.4A and B indicate the results. From Fig. 3.4A, it appears that the maximum of axial component velocity exists at $d/D_i' = 0.6$, which is the pipe axis, and the circumferential velocity component becomes maximum at $d/D_i' = 0.1$ in the peak side wall and vanishes within $d/D_i' = 0.4$–0.6. The flow angle of fluid at $d/D_i' = 0.1$ is almost equal to the spiral angle of the pipe. This result agrees well with the data obtained by the dye injection method as a matter of course.

By refereeing the experimental results of velocity profile, we can analyze the pressure loss of the spiral pipe approximately. The cross section includes three parts, which one of them is considered as a part of ellipse. We can analyze approximately the friction factor of laminar flow of a Newtonian fluid in the spiral pipe by applying the calculus of variations. Ito[6] made clear that the condition to minimize the energy dissipation for laminar flow equivalents to the equation which the inertia terms of N–S Equation of Eq. (2.19) become zero, and at the same time, an external

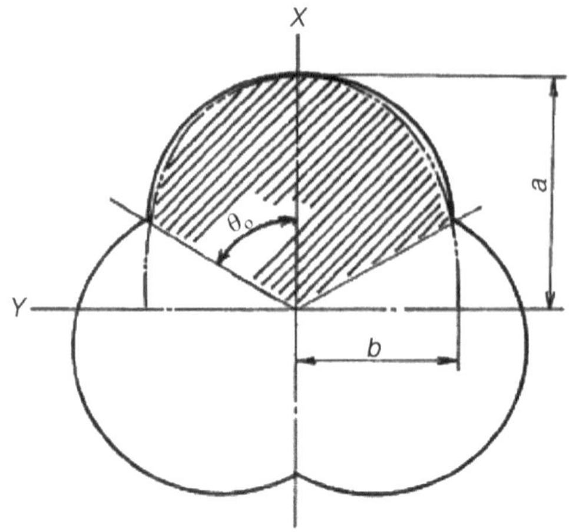

Figure 3.5 ***Calculated sectional area.***

force is derived from a potential or it becomes zero. Thus, we can obtain an approximate solution of the friction factor of a deformed pipe except a circular pipe by the method when it is very difficult to analyze directly from Eq. (2.19).

At first, we consider a Newtonian fluid in a straight spiral pipe having the same cross section, and then we assume that the flow in each cross section is a part of flow in a ellipse pipe because three cross section equivalent to each other. By applying the calculus of deviation, we can obtained relatively easily the friction factor of laminar flow[2] in a pipe with such as the cross section shown in Fig. 3.5.

It is given as following equation,

$$\lambda_s = \frac{8D_e^2}{3R_e} S\left(\frac{1}{ab\theta_B}\right)^2 \left\{\frac{(a^2+b^2)}{ab}\theta_B + \frac{(b^2-a^2)}{2ab}\sin 2\theta_B\right\} \tag{3.3}$$

where D_e and S area equivalent diameter of the spiral pipe, $D_e = 4M$ and the cross-section area. And M is the mean hydraulic depth. a and b are the major and minor axes of the approximate ellipse, θ_B is the angle which conforms to a part of the approximate ellipse as shown in Fig. 3.5.

Since the flow visualization results show that the fluids flow along the spiral wall, we can thereby be regarded as the friction factor of a spiral pipe with length, L should be equal to the value of the calculated straight pipe with the length of $(L/\cos\beta)$.

Therefore, it becomes

$$\lambda = \lambda_s / \cos\beta \tag{3.4}$$

The comparison between the experimental result and this analytical result are shown in Fig. 3.6. In the figure, the solid lines ① and ② are the solution of laminar flow and Prandtl-Kármán formula for a circular pipe, respectively.The friction factor of the spiral pipe increases with a decrease in the pitch, and it becomes parallel with the solid line ① in the laminar flow range.The transition point moves to the low Reynolds number region with a decrease in the pitch. Eq. (3.4) agrees almost with the experimental data.

According to the increase of R_e number, the flow becomes turbulent. Figs. 3.7 and 3.8 are indicated velocity profiles in the turbulent flow range. The figures attached to the right side of symbols in Fig. 3.7 identity the measured position in the pipe section as shown in Fig. 3.8, and the chain line in Fig. 3.7 is the one-seventh -power law velocity profile for a circular straight pipe The axial velocity becomes smaller than the one-seventh-power law velocity profile near the peak side wall of the spiral pipe with a decrease in the pitch.The circumferential velocity component vanishes near the pipe axis, and it has the maximum value at $d/D_i' = 0.1$.With a decreases in the pipe pitch, the maximum value increases, and it slightly increases in the downstream direction expect III-B type pipe.

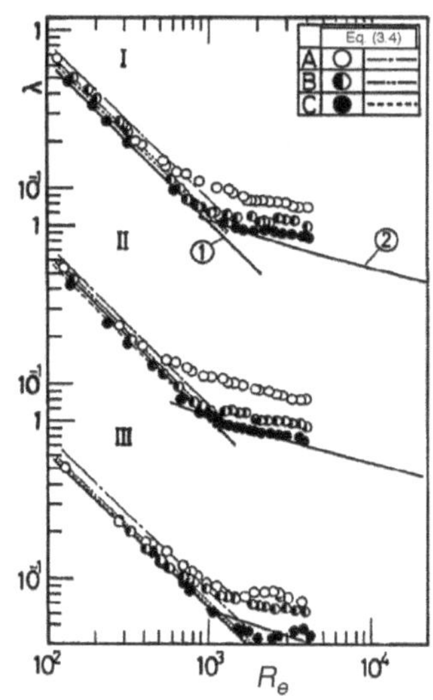

Figure 3.6 ***Comparison between experimental data and analytical results.***

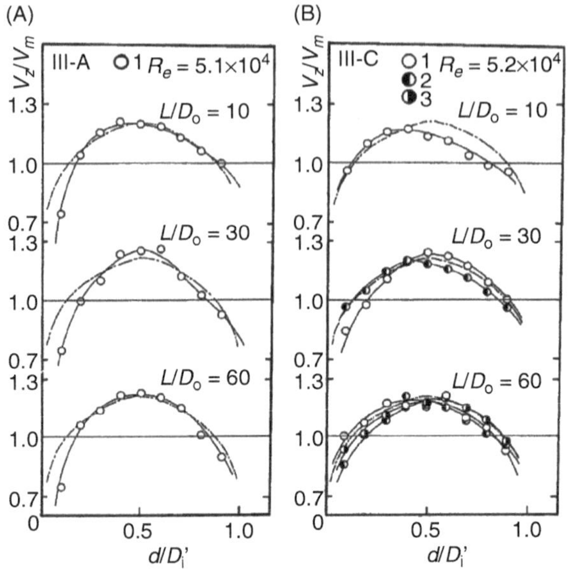

Figure 3.7 ***Axial velocity profiles in turbulent flow range.*** (A) III-A pipe. (B) III-C pipe.

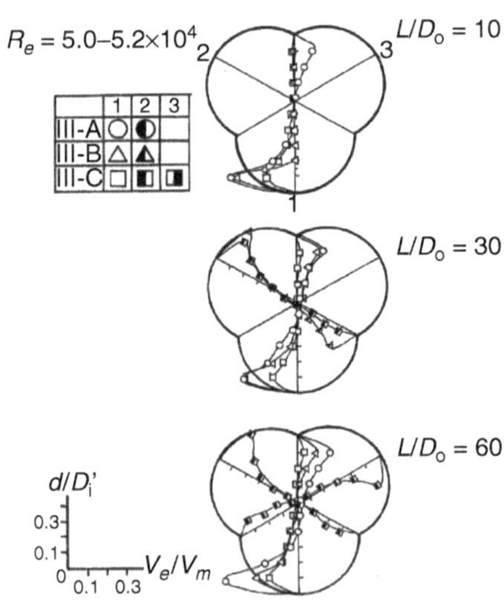

Figure 3.8 ***Circumferential velocity profiles in turbulent flow range.***

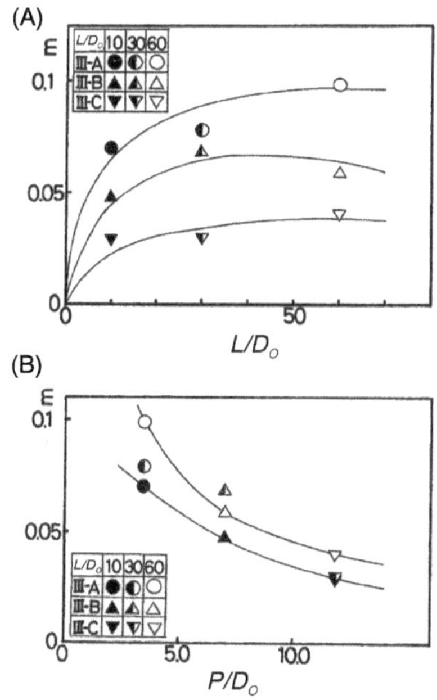

Figure 3.9 ***Variation of swirling number for inlet length and pitch.***

The turbulent mixing may be indicated using the swirling number as defined by following equation

$$m = \frac{\rho \int_A \upsilon_z \upsilon_\theta r\, dA}{\rho R \int_A \upsilon_z^2\, dA} \tag{3.5}$$

Fig. 3.9A and B show the variation of the swirling number against the nondimensional distance, L/D_0 and the nondimensional pitch, p/D_0, respectively. It shows that the value of the swirling number approaches to the constant value according to the increase of $L/D_{0.}$

On the other hand, the friction factor for turbulent flow is obtained experimentally by refereeing Colebrook formula for a rough circular pipe as follows,

$$\frac{1}{\sqrt{\lambda}} = -2.0 \log_{10}\left(B_0 + \frac{A_0}{R_e\sqrt{\lambda}}\right)$$

Hence, we assume that A_0 depends on the parameter (p/D_0) as the effect of the swirl flow, and B_0 is a function of $(\Delta d/D_0)$ as it is an equivalent parameter to the wall roughness of a circular pipe.

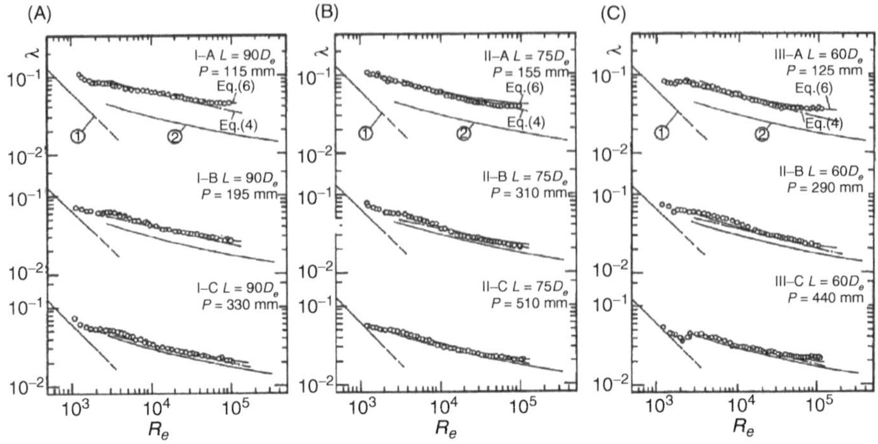

Figure 3.10 ***Comparison with experimental data.*** (A) Type I (B) Type II (C) Type III.

Consequently, the friction factor is given experimentally as follows,

$$\frac{1}{\sqrt{\lambda}} = -2.0\log_{10}\left[3.39\left\{\left(\frac{\Delta d}{D_0}\right)\tan\beta\right\}^{3.3} + \frac{18.05(\tan\beta)^{2.2}+2.51}{\mathrm{Re}\sqrt{\lambda}}\right] \quad (3.6)$$

$$\left(0 \le \Delta d/D_0 \langle 0.15, \quad 3\times10^3 \langle R_e \langle 10^5\right)$$

The comparison between Eq. (3.6) and the experimental data is shown in Fig. 3.10. The solid lines ① and ② in these figures, indicate the formula of laminar flow a circular pipe and Prandtl–Kármán formula, respectively. From these experimental results, it is seen that the Reynolds number dependences of friction factor λ are classified into two categories. They are A-region in which has gentle peak in the range of $2\times10^3 \le R_e \le 10^4$, and B-region in which is almost parallel with Blasius or Prandtl–Kármán formula, and after, it shows a similar tendency to that in the transition zone to the zone of complete turbulence for a circular pipe with the roughness. A-zone is observed over a wide the Reynolds number range in I~III-B type spiral pipes with $p/D_0 = 6-8$, but it has a limited range and the peak in small in I~II-A type spiral pipe. This has been reported on the friction factor of a corrugated pipe[7] in about same the Reynolds number range where the peak of the friction factor λ occurs in this range. The transition zone to complete turbulence remarkably appears in I type spiral pipe. From Fig. 3.4, it was seen that Eq. (3.5) are able to use as the formula of a Newtonian single-phase flow in turbulent flow range.

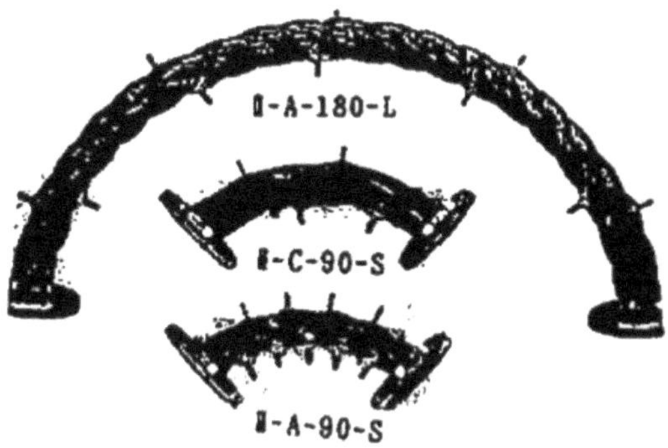

Figure 3.11 ***Spiral pipe bends.***

Other hand, an elbow or a bend is one of an important pipeline component in each industrial field. There are very few studies on flow in a spiral pipe bend although the study has been done about flow in a circular pipe[8] or a square duct bend.[9,10] Since the spiral pipe bend is fitted into the transport pipeline with the spiral pipe, it is necessary to clear the flow characteristics on the loss. Thus, the bend loss coefficient and the velocity profile are measured for the two types bends that the direction change angle, θ is 90° and 180°. Fig. 3.11 shows the spiral bend.[11] It is spiral pipe in the cross section shown in Fig. 3.2. In Fig. 3.11, it is marked the kind of pipe, I~A~III-C, the direction change angle, θ and the radios ratio, L or S. It is 12 and 18 in the radius ratio, (R/a).

Fig. 3.12 shows the measured axial velocity, v_z. The experimental data are indicated by the dimensionless expression that it divided by the velocity at the center axis, V_c. Data of a circular pipe bend in Fig. 3.12A show that the upward shift of the velocity is due to the effect of curved part, and it has a dimple in the center from upstream to downstream. Although it is agrees with the accepted experimental result for a circular pipe bend, the trend differs clearly according to the increase of the pitch in the case of the spiral pipe bend. As is evident from Fig. 3.12B and C, it changes convexly shape from upstream to downstream according to the increase of the pitch. The constant axial velocity lines obtained using an X type of hot wire probe, are indicated in Fig. 3.13. The numbers in the figure show that it is the ratio, (v_z/V_c). In the circular pipe bend shown in Fig. 3.13A fluids at the center axis, in which has large momentum, transfer to outer side by a centrifugal force, and it flows to the inside along to the wall. As the result, a sluggish speed fluid at the inside is pushed out of the center axis. However, the flow described above does, does not occur according to the increase of the pitch as shown in Fig. 3.13C.

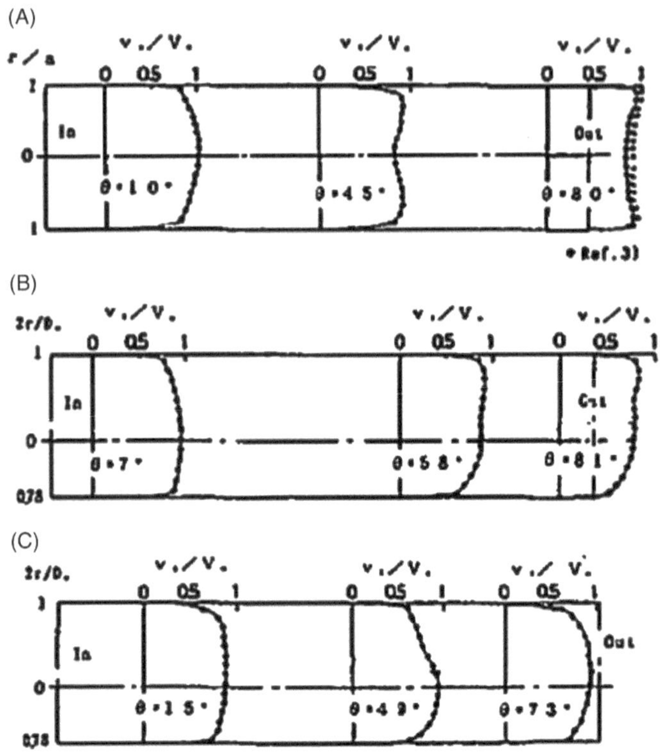

Figure 3.12 ***Axial velocity profile at*** $R_e = 5 \times 10^4$**.** (A) Circular pipe bend (III-90-S-L). (B) Spiral pipe bend (III-90-C-L). (C) Spiral pipe bend (III-90-A-L).

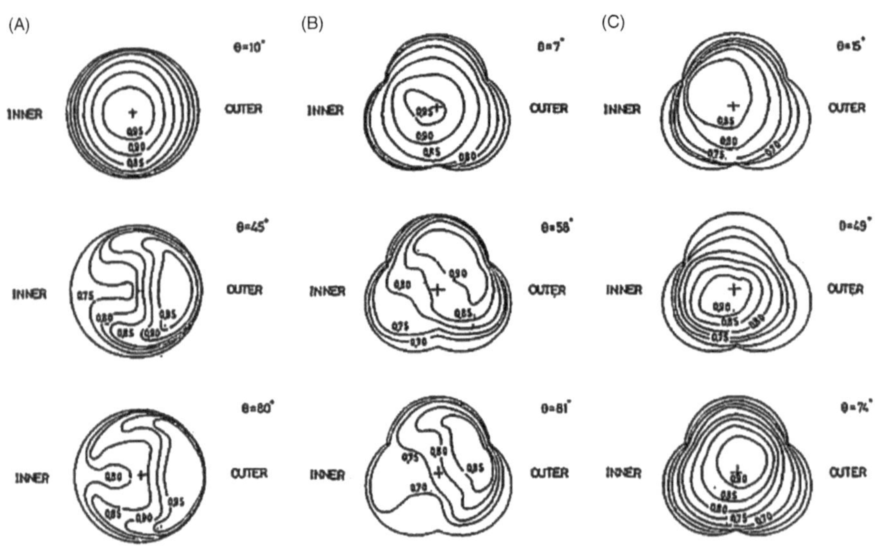

Figure 3.13 ***Constant axial velocity line at*** $R_e = 5 \times 10^4$**.** (A) Circular pipe bend (III-90-S-L). (B) Spiral pipe bend (III-90-C-L). (C) Spiral pipe bend (III-90-A-L).

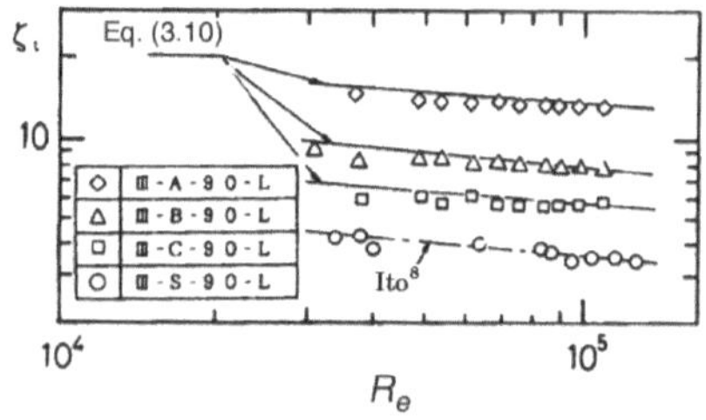

Figure 3.14 ***Total loss coefficient of bend.***

As a matter of course, it means that the fluids flow in a whirl along the spiral wall.

In general, it can be considered that the blockage of bend in gas–solid two-phase flow pipeline systems is occurred by unequal concentration of the solid particles that it occurs by the collisional slowing down of the solid particle or the migration to the wall by centrifugal force. Thus, we may be get off the blockage phenomenon in low velocity range by using the spiral bend because the maximum axial velocity position changes gyrating sort of movement.

The total loss coefficient of bend is defined as

$$\zeta_t = \frac{\Delta p}{\rho U^2 / 2} \tag{3.7}$$

where Δp is the pressure loss is given by the pressure distribution.[8] Fig. 3.14 shows the experimental results. Experimental data of a circular pipe bend agree well with the formula obtained by Ito. The results of the spiral pipe bend indicates the trend that the data shift to the upward parallel with an decrease of the pitch.

From the dimensional analysis, it is given as

$$\zeta_t = F(R_e, p/D_0, R/a, \theta$$

Experimental data are rearranged in Fig. 3.15 using non-dimensional parameter $[R_e(p/D_0)^6(R/a)^{-3}\theta^{-7}]$. The trend may be presented by the solid line by following equation,

$$\zeta_t = 0.85 \times \left[R_e^{-0.13} \left(p/D_0\right)^{-0.78} (R/a)^{-0.40}\, \theta^{-0.92} \right] \tag{3.8}$$

$$(R_e = 2\times10^4 - 10^5, \quad p/D_0 = 3.5 - 12.0)$$

The error span of Eq. (3.8) is shown in Fig. 3.16. Eq. (3.8) agrees within in the bound of ±20%.

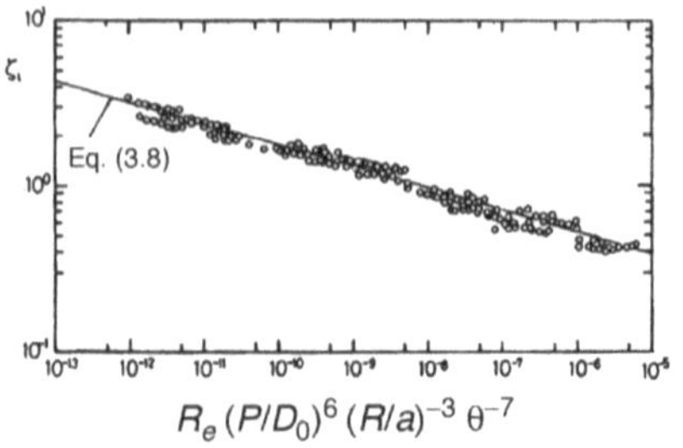

Figure 3.15 ***Total bend loss coefficient.***

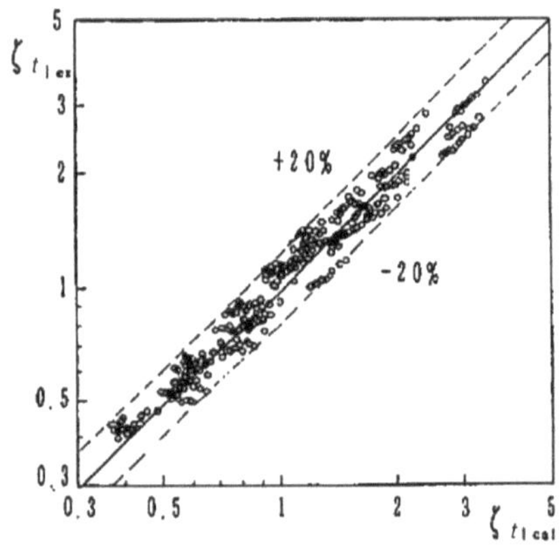

Figure 3.16 ***Comparison between Eq. (3.8) and the experimental data.***

Other hand, the loss based on the curved flow path is defined as the difference between the total bend loss and the pressure loss of an isometric pipe for the centerline, L_c. It is derived from

$$\varsigma_b = \varsigma_t - \lambda \frac{l_c}{D_e} \tag{3.9}$$

Since the friction factor of a spiral pipe is given by Eq. (3.6), the following equation is obtained experimentally,

$$\zeta_b = 0.135 \times \left[R_e^{-0.15} \left(p/D_0 \right)^{-0.88} (R/a)^{-0.44} \theta \right] \tag{3.10}$$

From Fig. 3.17, Eq. (3.10) agrees within in the bound of ±20%.

Based on the results, we can estimate the pressure loss of single phase flow in the pipeline system.

Solid–liquid two-phase flow is widely utilized in hydraulic transport of powered, low-grade coal, industrial waste products. It is expected that the

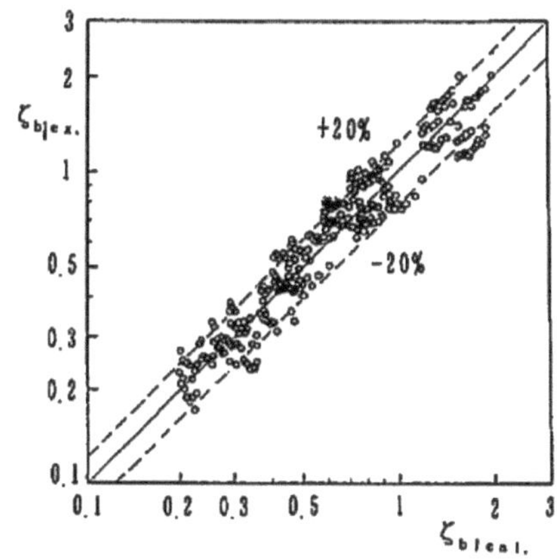

Figure 3.17 ***Comparison between Eq. (3.10) and the experimental data.***

utilization of the spiral pipe as transportation pipe, which puts the mixing effect of a swirling flow in the pipe into particle, can prevent settlement of solid particles and provide drag reduction.

Dimensions of test solid particle are listed in Table 3.1. The solid particles are a polypropylene pellet, an artificial light-weight aggregate and a crushed stone, respectively abbreviated as the P.P., A.L.A. and C.C. P.P. particles are cylindrical pellets about 1.8 mm in diameter and 3.2 mm in length. The particle size distributions of A.L.A. and C.S. particles are 2.5–5.5 mm. In Table 3.1, d_e is equal to the mean diameter of a sphere assumed by measuring the volume of one thousand solid particles, u_t is the terminal velocity of a particle in water obtained by using a rectangular tank (500 × 500 × 1600 mm), and the values of the terminal velocity, u_t in Table 3.1[3] represent the average of one hundred measured values.

Fig. 3.18A–C indicate the experimental data of pressure drop of heterogeneous flow[3] in the horizontal pipeline systems. The vertical axis and the horizontal axis represent the pressure drop per unit length,

Table 3.1 Dimensions of solid particles

Solid particles	**Mean diameter d_e(mm)**	**Density ρ_s (kg/m^3) × 10^{-3}**	**Terminal velocity μ_t (m/s)**	**Drag coefficient C_D (−)**
P.P.	2.52	1.44	0.079	2.01
A.L.A.	3.64	1.52	0.113	1.94
C.S.	3.49	2.52	0.209	1.59

A.L.A., Artificial lightweight aggregate; *C.S.*, crushed stone; *P.P.*, polypropylene pellet.

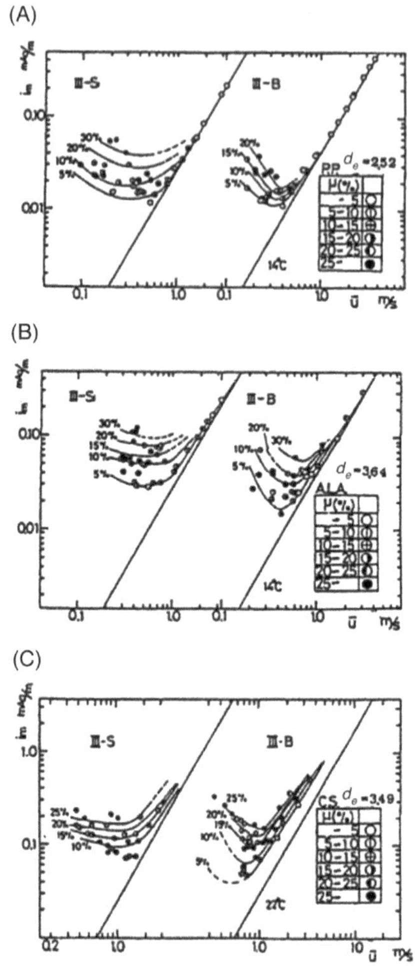

Figure 3.18 ***Pressure losses of horizontal pipes.*** (A) Polypropylene pellet. (B) Artificial lightweight aggregate. (C) Crushed stone.

$i = (\Delta p/\rho gL)$ and the mean velocity of mixtures, *U*. In Fig. 3.18, III-S means a circular pipe, which is 44.1 mm in a diameter, and III-B is the spiral pipe with $(p/D_0) \cong 7.0$ in Fig. 3.2. The straight lines represent the values of Blasius formula and Eq. (3.5) for a circular and a spiral pipe, respectively. The straight lines in these figures indicate the results for the flow of water alone, which were obtained by Blasius' formula and Eq. (3.6). The experimental data in these figures are shown under a certain concentration region, and the measured length for pressure loss in the spiral pipe is 3500 mm.

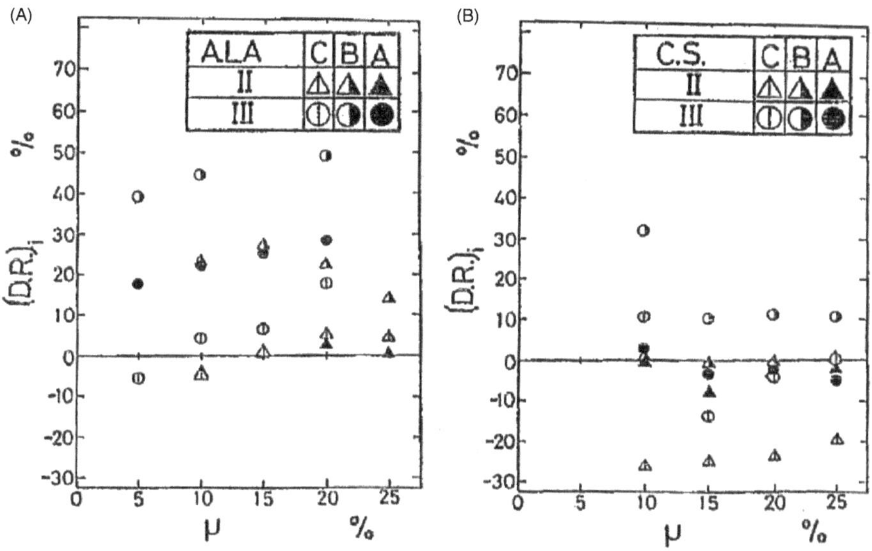

Figure 3.19 ***Variation of Drag reduction ratio for delivered mass fraction of solid particles.***

It can be seen from the figures that the pressure loss of the spiral pipe varies remarkably near the critical velocity region unlike the circular pipe, and the value of the critical velocity increases with the concentration I n the same manner as the data on circular pipes do. Since the value of the pressure loss in the spiral pipe is reduced compared with that of the circular pipe, Fig. 3.19A and B are arranged by using the drag reduction ratio defined by the following equation,

$$[D.R.] = \frac{(i_{cr})_s - (i_{cr})}{(i_{cr})_s} \times 100(\%) \tag{3.11}$$

where $(i_{cr})_s$ is the circular pipe value.

The maximum drag reduction occurs in a B-type spiral pipe of $p/D \div 7$, and its values in III-B type pipe are about 50% and 40% for A.L.A P.P. particles, respectively. Other hand, the value of the reduction ratio in II-C type pipe is small, and the pressure loss of C.S. particles is larger than that in circular pipe. The variation in the reduction ratio of the pressure loss to the delivered mass fraction of solid particles shows a similar tendency for P.P. and A.L.A. particles. These results suggest that a swirling flow insufficiently contributing to the floating of the particles causes additional increases in pressure losses due to the waved pipe wall, depending on the relation between the intensity of the swirling flow produced by the spiral pipe and the friction of solid particles against the pipe wall.

The pressure loss formula is obtained experimentally as follows,

$$\phi = KF_r^m \tag{3.12}$$

where

$$\phi = \frac{(i_m - i_w)}{i_w C}, \quad F_r = u\frac{u^{-2}}{gD_0(s-1)}\sqrt{C_D}$$

C, g, s, and C_D are the delivered volume fraction of solid, the gravitational acceleration, the specific gravity of solid and the drag coefficient of solid.

When discussing the flow of the solid–liquid mixture it is necessary to consider not only the pressure losses[3] but also the critical velocity, $\overline{u_{cr}}$ which is the velocity at the minimum value of the pressure loss. It is apparent from experimental results that the spiral pipe causes a reduction in $\overline{u_{cr}}$ and the reduction ratio increases with ($\Delta d/p$). The obtained values of the ratio are about 50% for A.L.A. or P.P. particles, and about 40% for C.S. particles in a II-A type pipe. Except for A.L.A. particles in a III-C type pipe, the value of $\overline{u_{cr}}$ is smaller than it is in circular pipes. Although it has been noted that value of $\overline{u_{cr}}$ obtained by differentiating the formula of the pressure loss depends on the still-unsatisfactory accuracy of the formula, $\overline{u_{cr}}$ is given by Eq. (3.12) as follows

$$\overline{u_{cr}} = \left\{k_1^7 - k_2^7\left(\frac{\Delta d}{p}\right)^{0.68}\right\}^{\frac{1}{3}}\left\{\frac{\sqrt{C_D}}{gD_0(s-1))}\right\}^{\frac{1}{2}} C^{\frac{1}{3}} \tag{3.13}$$

$$(0 \le \Delta d/p \le 3.74 \times 10^{-2})$$

where the experimental constants k_1' and k_2 are 92.5 and 745 for A.L.A. or C.S. particles, respectively, and $k_1 = 32.5$ and $k_2 = 263.5$ for P.P. particles. The comparison between the experimental data and Eq. (3.13) is shown in Fig. 3.20A and B the calculated value of Eq. (3.13) is represented on the abscissa and the experimental data on the ordinate. From Fig. 3.20A and B, it is seen that the experimental data are distributed around the correlation line and Eq. (3.13) agrees with the experimental data to an accuracy level of +15% to −20%. To clarify, the interaction of the particle suspension and the swirling flow, the experimental results with the acrylic resin spiral pipe are shown in Fig. 3.21A and B. Fig. 3.21A and B are photographs of

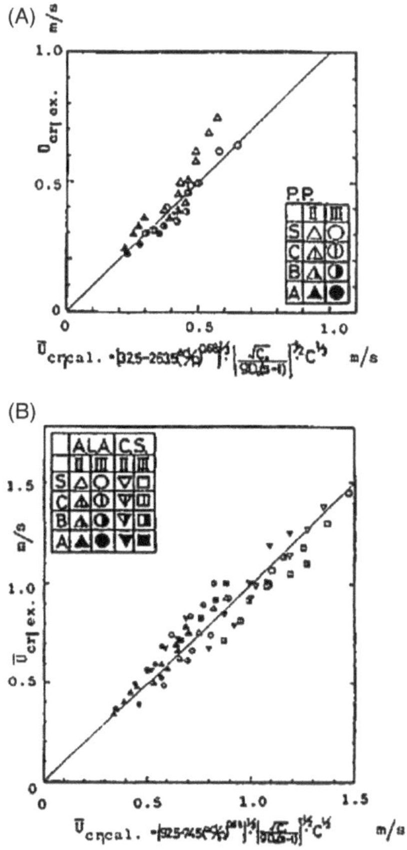

Figure 3.20 ***Correlation of experimental results and Eq. 3.13.***

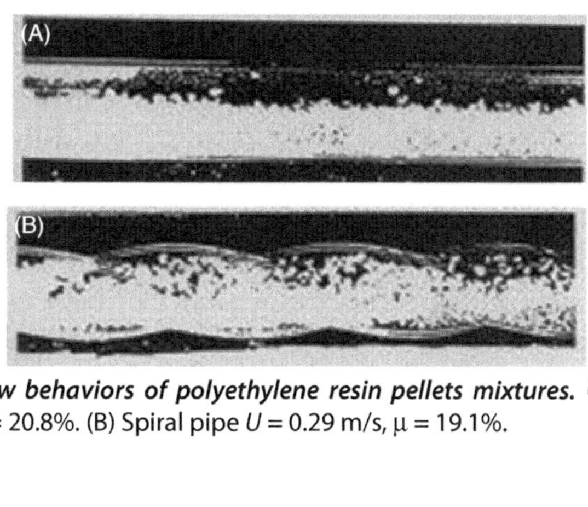

Figure 3.21 ***Flow behaviors of polyethylene resin pellets mixtures.*** (A) Circular pipe $U = 0.31$ m/s, $\mu = 20.8\%$. (B) Spiral pipe $U = 0.29$ m/s, $\mu = 19.1\%$.

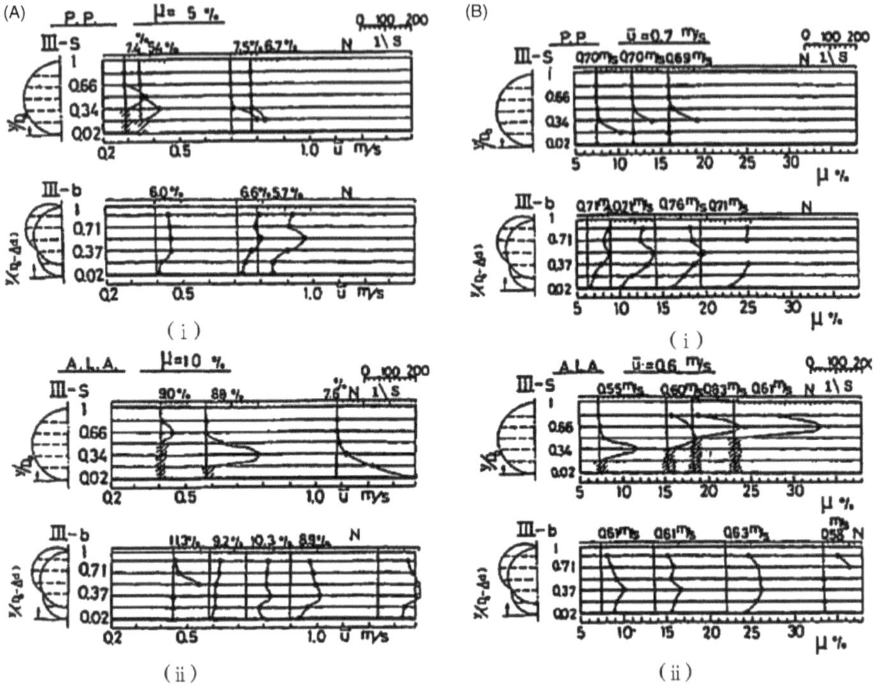

Figure 3.22 ***Particle concentration distribution.*** (A) Effect of the mean velocity. (B) Effect of the delivered mass fraction.

visualization in P.P. particles, and the flow condition corresponds to the nearly critical velocity range in both pipes. In Fig. 3.21A there is a settling bed near the pipe bottom, and only particles in the upper layer flow, but the particles in the spiral pipe are suspended in the upper layer and move with a sliding bed near the pipe bottom as shown in Fig. 3.21B.

The number of solid particles passing through the pipe section are measured by using a parallel beam and photo-transistor. The results of solid particles concentration in the pipe section are shown in Fig. 3.22A and B. The value of N represented on the ordinate of these figures is the number of particles flowing per unit time in the section, and it is obtained by reading the recorded pulse.

There are some cases where there is no distinction between the settling and sliding beds near the pipe bottom for the recorded data; thus, confirmation by visual inspection was required. In these figures, the sliding and settling beds are denoted by the symbol, Δ and the shadow area, respectively. According to the increase in mean velocity, the settling layer zone in the circular pipe decreases and the zone of water alone expends.

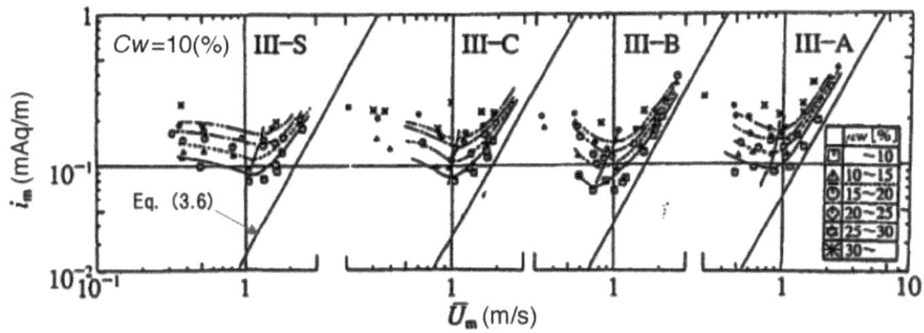

Figure 3.23 ***Pressure drops of crushed stone in 10% fly ash slurry.***

On the other hand, the settling bed in the spiral pipe does not exist within the flow range, and the solid particles still flow in the range of $y/(D_0-\Delta d)$ = 0.71–0.88. The distribution of particle concentration tends to be uniform with increasing u.

However, the value of $y/(D_0 - \Delta d)$ at the maximum value of N is 0.54 and 0.37 for P.P. and A.L.A. Particles, respectively, and are suspended by the swirling flow as compared with A.L.A. Particles as shown in Fig. 3.22B.

Currently, there is no fixed theory which compass the flow, drag reduction and pressure loss in the pipe of slurries containing coarse solid particles mixed with fly ash to that of water. It can be considered that inference is caused by solid particles interacting with the turbulence of the mixture flow or the pipe wall shear stresses of solid particles or by the characteristics of non-Newtonian fluids for the carrier fluids. Experiments were conducted with fluids concentration below the C_w = 10%–30% range in a III-type spiral pipe to clarify the drag reduction phenomenon.

Fig. 3.23 shows the typical pressure loss[12] for crushed stone-fly ash slurry flow in the horizontal spiral pipe. It is 10% in the case of crushed stone, which is $d_s = 3.49$ mm and $\rho_s = 2.52\times10^3$ kg/m^3 in the mean diameter and the density. The concentration of fly ash is $C_w = 10\%$. The straight lines are the results the flow of fly ash slurries alone which are evaluated by using the relative viscosity from Eq. (4.6) in Chapter 4. The curved lines were obtained from the extrapolation of the experimental data for the delivered weight fraction of crushed stones μ_w. The minimum value of the pressure loss is represented by the broken lines. It is very important to clearly understand the minimum pressure loss i_{mcr} and the mean velocity U_{mcr} because the hydraulic transport system is usually operated about critical velocity flow region. As shown in Fig. 3.23, the maximum reduction of i_{mcr} occurs in B-type spiral pipe of $(p/D_e) \cong 7.0$, and the i_{mcr} an A-type spiral pipe almost

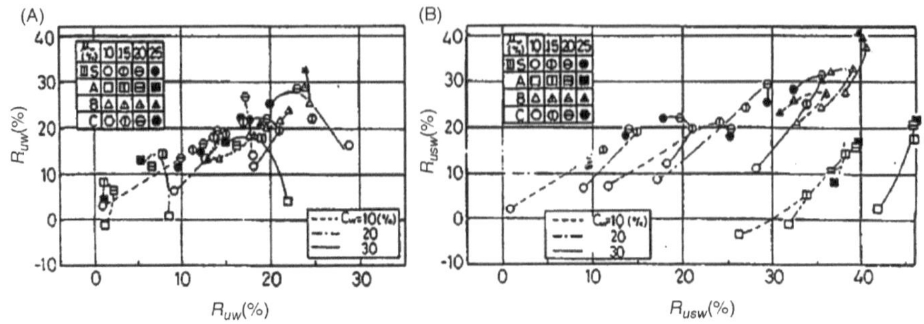

Figure 3.24 ***Drag reduction ratio.*** (A) R_{iw}–R_{uw}. (B) R_{iw}–R_{uw}.

equals that of a circular pipe. U_{mcr} of the spiral pipe decreases as its pitch increases.

It was shown that the maximum drag reduction ratio obtained for a B-type spiral pipe at the critical pressure loss was approximately 40% compared with that of circular pipe that carry water. Experimental data show that it is the B-type pipe in the most effective spiral pipe for the drag reduction of liquid–solid two-phase flow.

Fig. 3.24A and B, show the drag reduction ratios of i_{mcr} and U_{mcr} obtained from Fig. 3.23. They are defined as flowing equations

$$\left.\begin{aligned} R_{iw}(\%) &= \frac{i_{mcr}\big|_w - i_{mcr}}{i_{mcr}\big|_w} \times 100 \\ R_{uw}(\%) &= \frac{\overline{U}_{mcr}\big|_w - \overline{U}_{mcr}}{\overline{U}_{mcr}\big|_w} \times 100 \end{aligned}\right\} \tag{3.14}$$

where the subscript $|_w$ represents the crushed stone-water mixtures in each tested pipe. R_{iw} and R_{uw} in Fig. 3.24B are the reduction ratios defined by substituting the experimental data from the crushed stone-water mixtures in a circular pipe into Fig. 3.23. As shown in Fig. 3.24A, R_{iw} on a circular pipe is in the 18%–20% at μ_w = 15% and C_w = 20%–30%, and its value increases with the concentration of fly ash slurry. Other hand, R_{iw} of an A-type spiral pipe is smaller than that of circular pipe. The value of R_{iw} has a maximum in B-type spiral pipe at μ_w = 20% and C_w = 30%, and its value is about 40%. These results suggest that a spiral pipe of $(p/D_e) \cong 7.0$ for the hydraulic transport of a coarse solid particle-ultrafine particle mixtures flow is the most energy efficient. It is also interesting to note that the i_{mcr} of sand-water mixtures flowing in a spiral ribbed pipe[12] has a minimum for $(p/D_e) \cong 7.0-8$.

3.3 GAS–SOLID TWO-PHASE FLOW

As previously explained, the inertia force of solid particle is not negligible in gas–solid two-phase flow, since it is about 10^3 in the density ratio between the solid particle and gas. This is a major source of complexity in the flow pattern.

The flow visualization on perpendicular section of the pipe can reveal the effect of swirling flow on the movement of particles in a pipe. Fig. 3.25A and B shows the experimental apparatus.[13] The pipeline system and the measurement for the particle distribution in the section indicate in Fig. 3.25A and B, respectively. In Fig. 3.25B, the measuring part which is located at *A* in Fig. 3.25B, consists of a curved square pipe with a glass window ② Solid–gas mixtures flow into the transport pipe ③ from the pipeline ① through the measuring part. The measured cross section of the pipe ④ is painted black for shading except the slit width 10 mm. The measuring section is light up through a slit by stroboscopic light ⑤, and the particles are photographed by a camera ⑥ which is set on the coaxial point.

Fig. 3.26A and B are an example of the experimental results and the calculated results of the relative particle density distributions, respectively. Figure 3.26A indicates the results for wheat grains-air mixtures in a horizontal pipeline. In the figure, U_0 and m mean the air velocity and the loading ration, respectively.

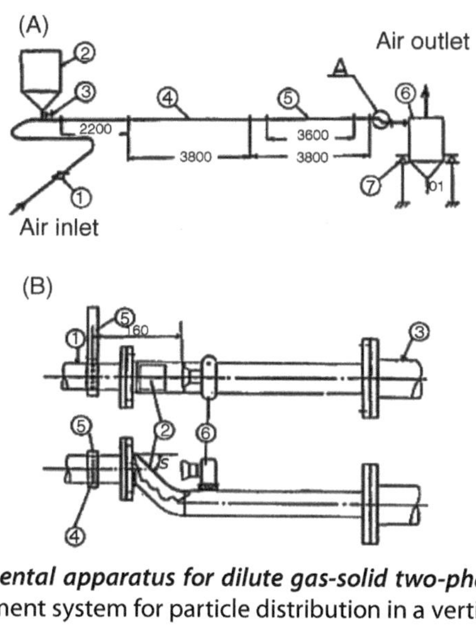

Figure 3.25 ***Experimental apparatus for dilute gas-solid two-phase flow.*** (A) Pipeline systems (B) Measurement system for particle distribution in a vertical section.

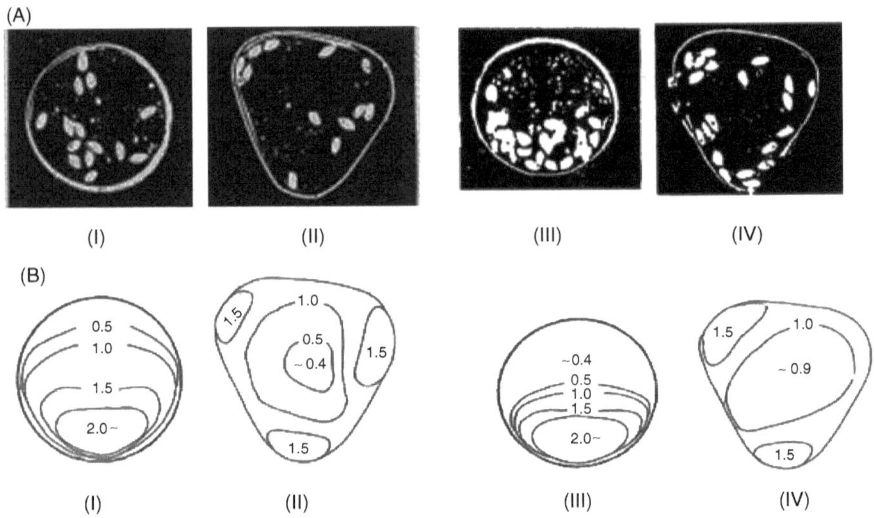

Figure 3.26 ***Distributions of wheat grains in the vertical section.*** (A) Photographs of wheat grains in pipes (i) *U*0 = 28.4 m/s, *m* = 7.4. (ii) *U*0 = 29.0 m/s, *m* = 7.7. (iii) *U*0 = 12.3 m/s, *m* = 6.3, (iv) *U*0 = 11.4 m/s, *m* = 11.2. (B) Calculated results of the relative particle density distributions.

As is evident from the comparison of the results between the circular pipe and the spiral pipe, it is seen that the particles in the spiral pipe do not deposit to the bottom of the pipe by the swirling flow in the low velocity range.

The application of gas–solid two-phase flow diverges into a pneumatic conveying system, a separating apparatus for particulate and a pulverized coal combustion apparatus. Since the pressure loss of a pipeline relates directly to the power for operating system, and it depends on the flow-pattern, it is necessary to grasp the relationship between the flow-pattern and the pressure loss. The latest technique trends of pneumatic conveying system are dense phase and low conveying velocity to protect the wear of pipe wall. And the system is expanded for object to be transported particles, which were not targeted for pneumatic conveying system conventionally. For example, we will be not able to apply the conventional pneumatic conveying system for the high concentrations of potassium alum pellets, feldspar powders, and sintered powders for steel, because the blockage occurs in a circular pipeline. However, the application of the spiral pipe to the system may be a feasible technology to avoid the blockage of solid particles as sown by the flow visualization experiments.

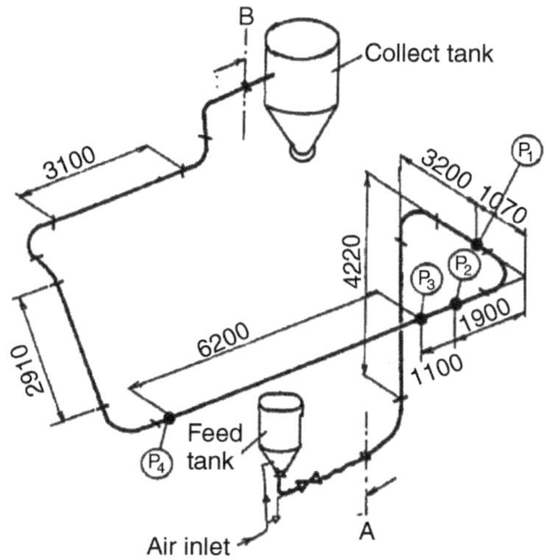

Figure 3.27 ***Pipeline system for dense gas-solid two phase flow.***

Fig 3.27 indicates the pipeline systems of the experimental apparatus.[14] Test particles in a feed tank flow in the pipeline having total length 33 m, and are corrected into a collect tank. Each pipeline of a circular pipe and a spiral pipe made of stainless steel is laid parallel. The diameter of a circular pipe is D_e= 54.9 mm, and the hydraulic diameter of a spiral pipe is D_e = 51.4 mm. The spiral pipe is equivalent to Type-B having $(p/D_0) \cong 7.0$ in Fig. 3.2. The pressure losses of a horizontal pipe are obtained by measuring the pressure deference $(P_3–P_4)$.

The conveyed bulk materials are listed in Table 3.2. The mean diameter of ABS pellets is given as volume-based particle size since it is a cube 2.5 mm on each side. The diameter of other bulk materials is decided by D_{50}

Table 3.2 Dimensions of conveyed bulk materials

Material	Mean diameter d_e(μm)	Density ρ_s (kg/m³)	Floating velocity V_{f^*} (m/s)	Loading ratio *m* (–)
ABS pellets	3102	1040	8.5	27–51
Potassium alum pellets	1730	1750	6.9	47–175
Feldspar powder	420	2570	2.0	82–144
Silica sand	350	2680	2.5	99–568
PVC powder	180	1420	0.6	42–97
Sintered powder For steel	53	5130	0.5	98–235
Calcium carbonate powder	5.9	2710	0.003	80–150

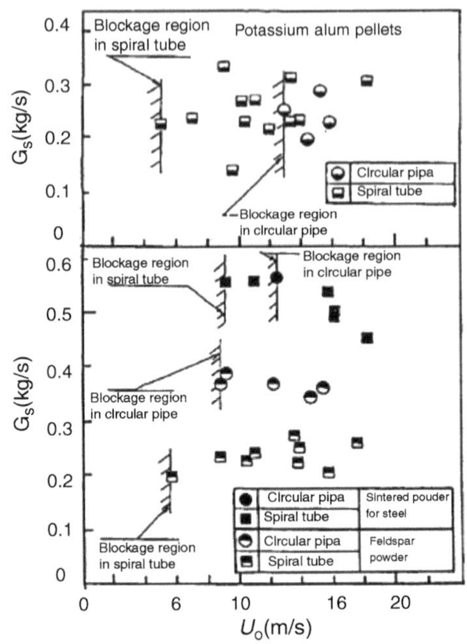

Figure 3.28 ***Comparison of blockage regions.***

particle size distribution. In other words, it physically represents that each number of particles greater or smaller than such value takes account of 50% of the total particles number. The floating velocity is calculated from the free-falling velocity of a sphere.

The relationship between the total transport amount G_s and the apparent air velocity reference atmosphere pressure in these pipes U_0, is indicated in Fig. 3.28. The materials are potassium alum pellets, sintered powder for steel, and feldspar powder. It is known that the blockage occurs easy relatively in the conventional pneumatic conveying system. In the figure, the indicated with hatched indicates that the blockage occurs in the velocity where is lower than it. In the case of potassium alum pellets, although the blockage occurs in the range of $U_0 < 13$ m/s, it occurs in the range of $U_0 < 5$ m/s. The critical blockage velocity of the spiral pipeline system decreases in comparison with that of the circular pipeline system.

Fig. 3.29A and B show the pressure fluctuation of calcium carbonate powder and ABS pellets, respectively. An ABS pellet has a cubic shape, 2.5 mm on a side. The number ①–④ in these figures indicate the pressure

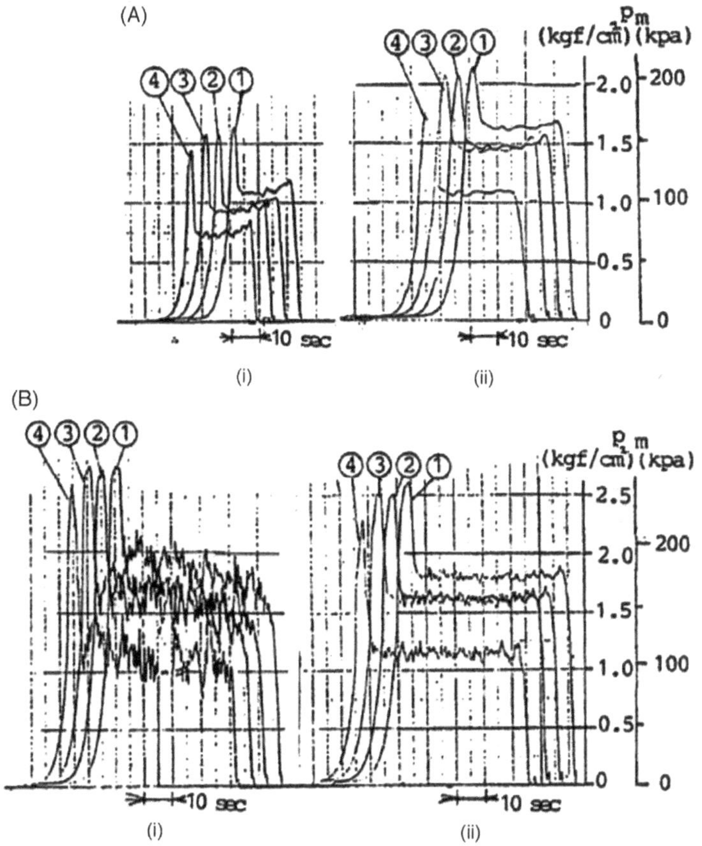

Figure 3.29 ***Pressure fluctuation of ABS pellets mixtures in a horizontal pipe.*** (A) Calcium carbonate powder (i) Circular pipe, $U_0 = 17.5$ m/s, $m = 94.3$; (ii) Spiral pipe, $U_0 = 19.3$ m/s, $m = 64.6$. (B) ABS pellets (i) Circular pipe, $U_0 = 16.7$ m/s, $m = 41.1$; (ii) Spiral pipe, $U_0 = 18.1$ m/s, $m = 41$.

monitoring position shown in Fig. 3.27. Once the pressure increases at the close of the transport and after it vanishes. The variation in pressure fluctuation is little for the spiral pipeline system in comparison that of the circular pipeline system in each case of solid particles. That is, the stable condition of transport is maintained in the spiral pipeline.

The pressure loss of gas–solid two-phase flow in a horizontal pipe is given by Eq. (1.1). By definition for Δp_f and Δp_a as follows,

$$\left.\begin{aligned} \Delta p_f &= \lambda_a \frac{L}{D_e}\frac{\rho_a U_a^2}{2} \\ \Delta p_a &= m\lambda_s \frac{L}{D_e}\frac{\rho_a U_a^2}{2} \end{aligned}\right\} \tag{3.15}$$

where ρ_a and U_a are the density and the mean velocity of air, and λ_a is given by Eq. (3.6). m and λ_s are the loading ratio and the friction factor of the gas–solid two-phase flow.

In Eq. (3.16), (λ_s/λ_a) is given by using the Froude numbers, $F_D = w_f/\sqrt{gD_e}$ and $F_r = U_a/\sqrt{gD_e}$ as follows,

$$\frac{\lambda_s}{\lambda_a} = 2.2\times 10^5 \left(d_e/D_e\right)^{1.7} \frac{1}{F_D\left(F_r - F_D\right)} \tag{3.16}$$

The result in the correlation of Eq. (3.16) are indicated in Fig. 3.30 The experimental data of ABS pellet, PVC powder and sintered powder for steel, but does not fit with Eq. (3.16). By floating velocity and characteristic length of solid particles, it doesn't help explain for the pressure loss in gas–solid two-phase flow.

The pressure loss of bend Δp_{bm} is obtained by the pressure difference between P_1 and P_3 in Fig. 3.27.

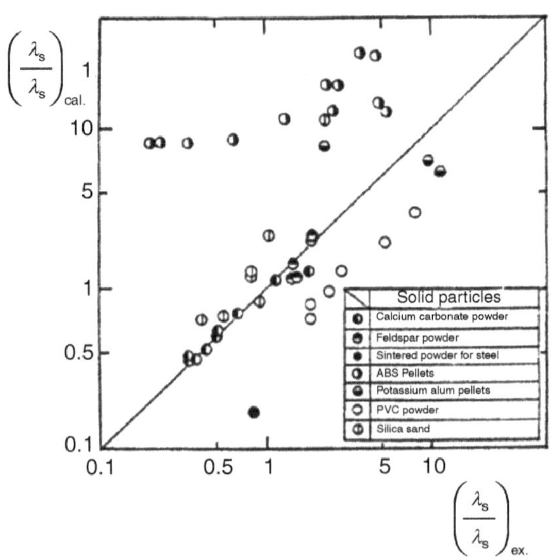

Figure 3.30 ***Correlation between the experimental results and Eq. (3.16).***

$$\Delta p_{bm} = \Delta p_{31} - \Delta p_{m} = \Delta p_{bl} + \Delta p_{bss} \tag{3.17}$$

where Δp_{31} and Δp_m are the pressure difference between P_1 and P_3 and the total pressure loss of the straight pipe in the section. By assuming Δp_{bm} is given the sum the pressure loss of carrier gas only, Δp_{bl} and the additional pressure loss of two-phase flow, Δp_{bs}, they are expressed as

$$\left.\begin{aligned} \Delta p_{bl} &= \zeta_l \frac{\rho_a U_a^2}{2} \\ \Delta p_{bs} &= \zeta_k \frac{\rho_m V^2}{2} \end{aligned}\right\} \tag{3.18}$$

ρ_a and U_a are the density and the mean velocity of air, respectively. ρ_m is the density of the particle group in unit volume of the pipeline, and V is the mean velocity of the particle group. The bend coefficient of air, ς_i is given as

$$\varsigma_i = 335\, R_e^{-0.535} (10^4 \langle R_e \langle 10^5)$$

Fig. 3.31A and B shows the experimental results of the additional bend loss coefficient for two-phase flow, ζ_k. We can express the results for the circular pipe bend by the following equation as shown in Fig. 3.31A,

$$\zeta_k = 35 F_r^{-1.4} \tag{3.19}$$

$$(R/D_e = 16.4,\ \ 27\langle m \langle 235)$$

However, Eq. (3.19) does not fit with the experimental data of the spiral pipe bend. Experimental data decrease in comparison with Eq. (3.19).

The ratio between the additional bend loss coefficient of the circular pipe bend, ζ_k and that of the spiral pipe bend, ζ_{ks} are shown in Fig. 3.32A and B. Since the value of the ratio is smaller than 1.0 in cases of ABS pellets, sintered powder for steel and potassium alum pellets, we can confirm the drag reduction phenomenon in these cases. The usefulness of the spiral pipeline system for high concentration transport was also demonstrated by other pipeline system in 60 m length.[14] As pointed out above, the solid particles of interference to the pipe wall or the sudden turn of the particle velocity are the dominating factor for the total pressure loss of a bend. Thus, we can hold promise not only the prevention of blockage but the drag reduction in the pressure loss.

On the other hand, the drag reduction of dilute gas–solid particle suspension will be a phenomenon worthy of note. Although this has been

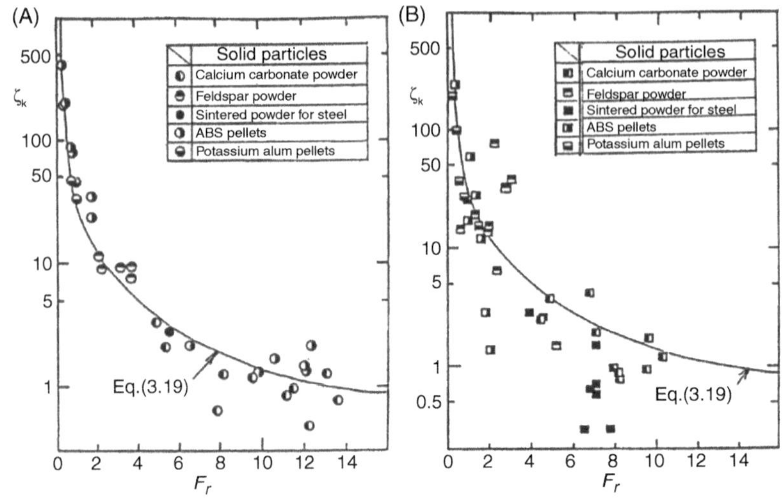

Figure 3.31 ***Additional bend loss coefficient.*** (A) Circular pipe bend. (B) Spiral pipe bend.

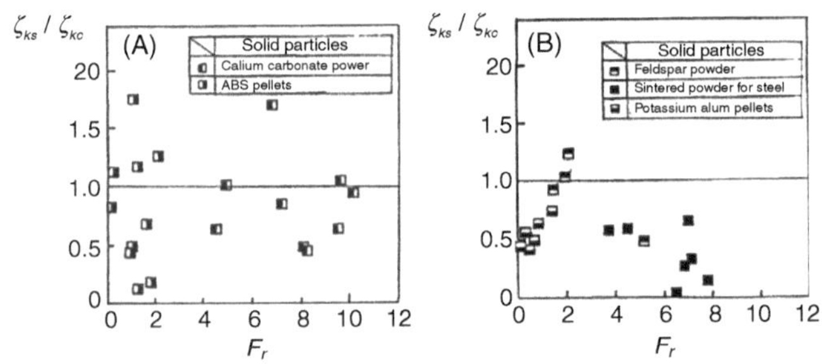

Figure 3.32 ***Ratio of additional bend loss coefficients.***

identified for a long time, one cannot necessarily conclude since the condition of the published paper on the drag reduction of gas–solid dilute suspensions does not agree only on the drag reduction ratio but also on the experimental flow conditions as pointed Pfeffer and Kane.[15]

Table 3.3 are summarized on some experimental results[16–21] for a horizontal pipe by refereeing the paper. It is clarified experimentally that the drag reduction occurs for dilute air suspension of grass having mean diameter of 10–30 μm in a horizontal pipe. The flow conditions are the pipe diameter range 20–70 mm, the Reynolds number range of 2.8×10^4–1×10^5 and the loading ratio range of 0.1–3, respectively.

Table 3.3 Drag reduction of dilute gas–solid two phase flow

References	Pipe material	Mean diameter d_e (μm)	Pipe diameter D (mm)	Loading ratio m (−)	Reynolds number R_e (-)	Maximum drag reduction ratio (%)
Rossetti and Pfefer[16]	Glass	10	22.1	~1.5	2.8×10^4	25
		20	22.1	~2.0	2.6×10^4	20
		30	22.1	2.2	2.6×10^4	18
Kane et al.[17]	Glass	15	22.1	~0.9	$(1.1–2.3) \times 10^4$	
		2 1	22.1	~1.0		10
Soo and Trezech[18]	Magnesia	30	12.7	~1.0	$(1.3–3.0) \times 10^4$	5
Clark et al.[19]	Glass	1050	25.4	1.5–5.0	(1.2–6.8)	-
Richardson and McLeman[20]	Perpex	1524	25.4	~1.0	$(1.2–7.0) \times 10^4$	-
Boyce and Blick[21]	Glass	5–1680	68.9	~3.0	$(1.0–6.3) \times 10^4$	10

Data of Coughran[22] for air suspensions of spherical particles and two types of fibrous particles, all having mean diameter of between 8 and 20 μm show that drag reduction occurs in air-spherical suspensions however uniform fibers with an aspect ratio of 75 give no drag change, while irregular or random cut fibers with a mean aspect ratio of 31 caused drag or pressure drop increases. As the inference factor on drag reduction, although we can consider decrease of gas viscosity, electrostatic charging, turbulence modification and relaxation, the difficulty remains in the analysis for the flow behavior of dilute particle suspension yet. Difference between liquid and gas suspensions in drag reduction finding might help to clarify this mechanism.

When dilute gas–solid suspensions flow in a plexiglass pipe, the electrostatic charge is generated between the inner wall of pipe and the glass particles in suspensions and they become the charged particle. Because the drag reduction occurs in the case, it can be considered that the electrostatic charging is one of the dominant contributors for the reduction. Future efforts in the study on dilute gas–solid suspensions should concentrate on the basic mechanism for the observed drag reduction phenomenon.

3.4 GAS–LIQUID TWO-PHASE FLOW

Gas–liquid two-phase flows occur by not only cavitation or gas blowing but the pipe line of geothermal power generation or various kinds of chemical plants. Gas bubbles in a flowing liquid decrease the pressure response velocity or the heat transfer coefficient, and the existence of gas and the liquid relative velocity cause an energy loss. The most common way of gas bubbles separation is the application of a swirling flow. The equipment is known as a liquid cyclone in general. However, the flow direction changes and the pressure loss increase by set up it in the pipeline system. Thus, if we can converge these flowing gas bubbles with liquid in the pipeline to the radial direction, it may be solved these problems. Since the swirling flow occurs in a spiral pipe as shown in Fig. 3.2, it may be applying to the separation of air bubbles in the pipeline.

At first, we clarify the behavior of a bubble[23] in the spiral pipe listed in Table 3.4 by a numerical simulation. The physical constants of fluids is given in Table 3.5. When the circumference component of the velocity profile is assumed as follows

Table 3.4 Dimensions of spiral pipe

Pipe	r_D (mm)	r_o (mm)	r_i (mm)	Δd (mm)	P (mm)
III-A	25.4	22.3	18.5	5.0	15.5

Table 3.5 Physical constants of liquid

	Density ρ_L (kg/m³)	Surface tension σ (N/m)	Viscosity μ (Pa·s)
Tap water	999.7	0.072	1.307×10^{-3}
Turbine oil	857.0	0.040	32.50×10^{-3}

1. It is gives by the combination of a velocity profile in the section at $30° \leq \theta \leq 150°$.
2. It increases linearity from the center to the maximum velocity position and after decreases linearity with an increase of the radius.
3. The maximum velocity is related to the mean velocity, *Um* and they are given as $U_{mm} = 0.4U_m$ and $U_{mi} = 0.2U_m$ that U_{mm} and U_{mi} are the maximum velocity at $r= r_m$ and $r= r_i$, respectively.
   Thus, the velocity profiles become

$$U_\theta = \left(U_{mm} - H\left[\theta - \frac{\pi}{2}\right]\right)\frac{r}{r_m}\left(0 \leq r \leq r_m\right) \tag{3.20}$$

$$U_\theta = \left(U_{mm} - H\left[\theta - \frac{\pi}{2}\right]\right)\frac{R_m - r}{R_m - r_m}\left(r_m \leq r \leq R_m\right) \tag{3.21}$$

where H is the mean value which is defined as $H = 3\left(U_{mm} - U_{mi}\right)/\pi$.

The radial component of the velocity profile is obtained by using the continuity equations for each range,

$$U_r = -\frac{H}{r_m}\frac{r}{2} \qquad \left(0 \leq r \leq r_m\right) \tag{3.22}$$

$$U_r = \frac{H}{r}\frac{r_m}{2} + \frac{H}{r}\frac{r - r_m}{R_m - r_m}\left(R_m - \frac{r - r_m}{2}\right) \qquad \left(r_m \leq r \leq R_m\right) \tag{3.23}$$

As one example, Fig. 3.33 shows the velocity vector of $U_m = 2.5\,\text{m/s}$. In the figure, the dotted line represents the position of the maximum velocity.

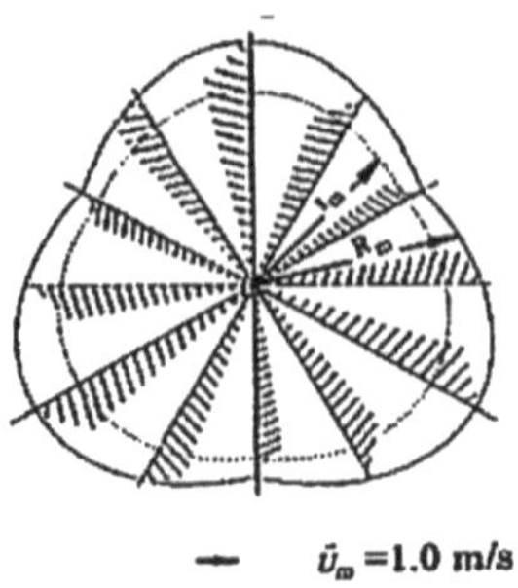

Figure 3.33 ***Assumed velocity profiles water in the spiral pipe.***

Equation of motion for a air bubble in the spiral pipe is written

$$\left(m_g - \alpha m_L\right)\frac{d\boldsymbol{U}_g}{dt} = \boldsymbol{F}_1 + \boldsymbol{F}_2 + F_3 \tag{3.24}$$

where m_g and Ug are the mass and the velocity of a bubble. α and mL are the virtual mass coefficient[24] and the mass of liquid which equals to volume of a bubble, respectively. In the case of a sphere bubble, the value of α is used 0.5.

In Eq. (3.24), F_1, F_2, and F_3 are the buoyant force, the drag and the force that is caused by the pressure gradient, and they are deduced

$$\boldsymbol{F}_1 = \left(m_L - m_g\right)\boldsymbol{g}$$

$$\boldsymbol{F}_2 = \frac{\pi a^2}{2} C_d \rho_L \left(\boldsymbol{U}_g - \boldsymbol{U}_L\right)\left[\boldsymbol{U}_g - \boldsymbol{U}_L\right]$$

$$\boldsymbol{F}_3 = \rho_L V g \left(\frac{DU_L}{Dt} - \frac{\mu}{\rho_L}\nabla^2 U_L\right)$$

C_D is the drag coefficient of a bubble[25] given by

$$C_D = max\left[\frac{24}{R_e}\left(1 + 0.15\ R_e^{0.687}\right),\ \frac{8}{3}\frac{E_0}{E_0 + 4}\right] \tag{3.26}$$

E_0 is Eötvös number defined as $E_0 = g\left(\varrho_L - \rho_g\right)d^2/\sigma$, where d and σ are the bubble diameter and the surface tension, respectively. R_e is Reynolds number that $R_{er} = 2a\left(dr/dt\right)\varrho/\mu$ and$R_{e\theta} = 2ar\left(\Omega - d\theta/dt\right)\varrho/\mu$ are given for the radial and the circumference direction. From Eq. (3.24) which is calculated by using Eqs. (3.20)–(3.23) and Eqs. (3.25)–(3.26), the bubble trajectory is calculated by Runge-Kutta method. It is $d = 0.7 - 3.0\,\text{mm}$ and

$U_m = 0.5 = 2.51\,\text{m/s}$ in the diameter and the mean velocity for the calculation range. By assuming the velocity of bubble equals to that of liquid at the start, the starting position is changed in the range of $r = 0.2R_m - 0.9R_m$. Although we calculate it for and , the result for is shown because the same result is obtained.

The calculation results of a air bubble trajectories are shown in Fig. 3.34A–D. Fig. 3.34A and B show the effect of the diameter and the mean velocity on the trajectory, and (R_c/R_m) in the figures is the nondimensional critical convergence radius obtained by the calculation. It means that the bubble converge if the initial position is $r \langle R_c$. In this calculation, the rebound of the bubble at the pipe wall is ignored, and if it hit the wall.

The results give the two convergence patterns which are converging at one point as shown in Fig. 3.34A(i) and B(ii), and the other pattern is moving along an elliptic orbit as shown in Fig. 3.34A(i)–(ii) and B(i)–(ii). From Fig. 3.34A and B, it shown that the bubble in water converges to the center of the pipe with an increase of the mean velocity or a decrease of the diameter, and the critical convergence radius, R_c increases.

Fig. 3.34C and D show the results for a air bubble in turbine oil. We can see that the bubble converges in one point of the section although the pass line is long in comparison with that in water. It can be considered that the influence of the drag acting on the bubble becomes dominant since the buoyant force decreases by the decrease of the density of liquid. In other words, it is required many times to converge the bubble in the pipe section.

From the calculation results for water, we can obtain the relationship between the critical convergence radius and the diameter of bubble. Fig. 3.35 shows the results. The results of turbine oil are presented by the dotted lines in comparison. The critical convergence radius decrease with an increase of the radius in the range $U_m \leqq 1.76$ m/s and the radius becomes constant in the range of $U_m \geq 2.01\,\text{m/s}$. By using π theorem described in Chapter 1, the result are rearranged using the nondimensional parameters as follows,

$$\left(\frac{R_c}{R_m}\right) = f\left(M, E_0, R_e\right) \tag{3.27}$$

where M, E_0, and R_e are Morton number, $M = g\left(\rho_L - \rho_g\right)\mu^4/\left(\rho_L^2\sigma^3\right)$, Eötvös number, and Reynolds number. The analysis of the calculation result give the following equations.

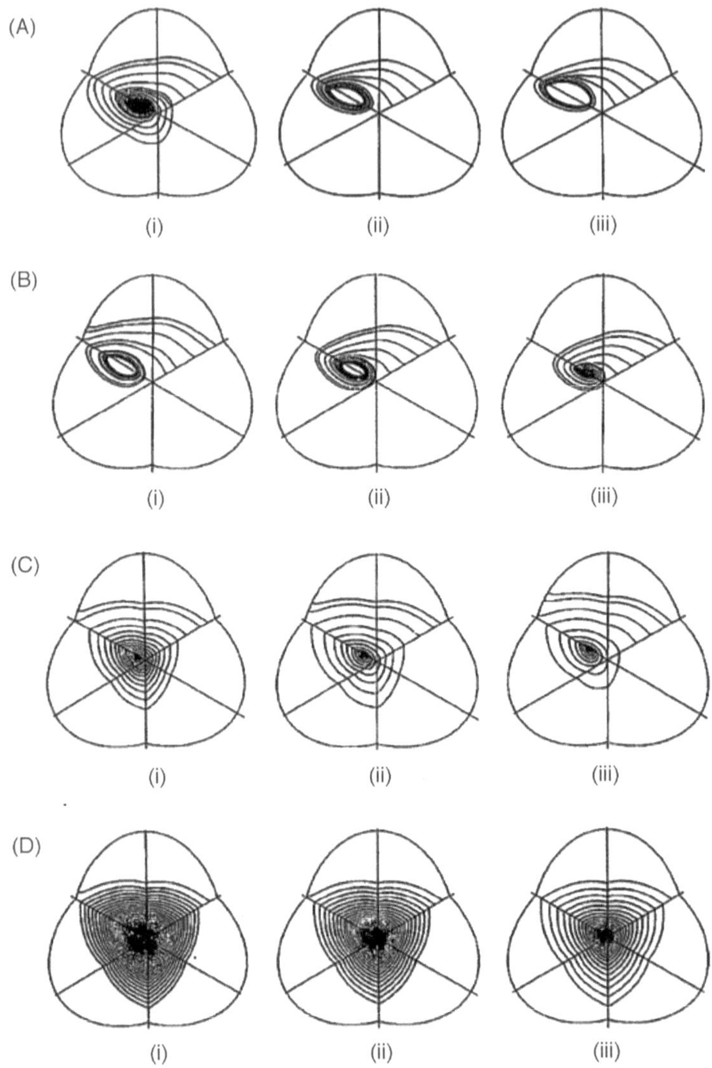

Figure 3.34 ***Numerical simulation results for the trajectory of air bubbles.*** (A) Trajectories of air bubbles in tap water at $U_m = 1.76$ m/s, (i) $d = 1.0$ mm ($R_c/R_m = 0.90$); (ii) $d = 2.0$ mm ($R_c/R_m = 0.84$); (iii) $d = 2.5$ mm ($R_c/R_m = 0.81$). (B) Trajectories of air bubbles ($d = 1.5$ mm) in tap water, (i) $U_m = 1.51$ m/s ($R_c/R_m = 0.36$); (ii) $U_m = 2.01$ m/s ($R_c/R_m = 0.90$); (iii) $U_m = 2.51$ m/s ($R_c/R_m = 0.90$). (C) Trajectories of air bubbles in turbine oil at $U_m = 1.01$ m/s, (i) $d = 1.5$ mm ($R_c/R_m = 0.79$); (ii) $d = 2.0$ mm ($R_c/R_m = 0.69$); (iii) $d = 2.5$ mm ($R_c/R_m = 0.54$). (D) Trajectories of air bubbles ($d = 1.0$ mm) in turbine oil, (i) $U_m = 0.76$ m/s ($R_c/R_m = 0.84$); (ii) $U_m = 1.01$ m/s ($R_c/R_m = 0.89$); (iii) $U_m = 1.51$ m/s ($R_c/R_m = 0.90$).

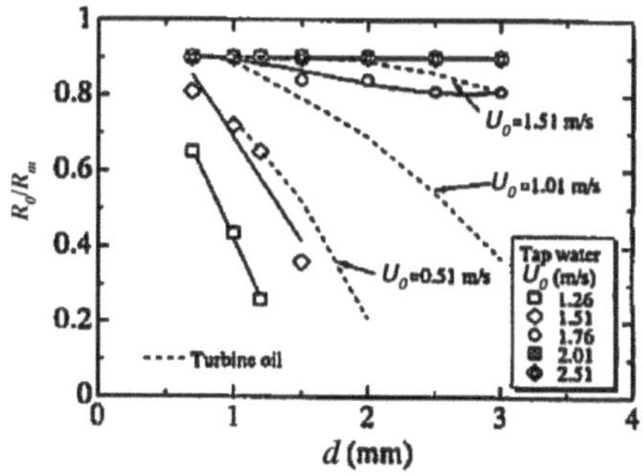

Figure 3.35 ***Convergence radius of air bubbles in tap water.***

For bubble in water;

$$\left.\begin{aligned}&\left(\frac{R_c}{R_m}\right)=2.81\times10^{-7}\times\left(M^{1.68}E_0^{-0.17}R_e^{5.04}\right)\\&\left(3.5\times10^3\le\left(ME_0^{-0.1}R_e^3\right)<7.5\times10^3\right)\\&(0.7\,\text{mm}\le d\le3.0\,\text{mm})\\&\left(\frac{R_c}{R_m}\right)=0.9\\&\left(7.5\times10^3\le\left(ME_0^{-0.1}R_e^3\right)<4.8\times10^4\right)\\&(0.7\,\text{mm}\le d\le3.0\,\text{mm})\end{aligned}\right\}\tag{3.28}$$

For bubble in turbine oil;

$$\begin{aligned}&\left(\frac{R_c}{R_m}\right)=1.46\times10^{-3}\times\left(M^{0.63}E_0^{-0.94}R_e^{1.88}\right)\\&\left(2.6\times10^3\le\left(ME_0^{-1.5}R_e^3\right)<2.9\times10^4\right)\\&(1.0\,\text{mm}\le d\le3.0\,\text{mm})\\&\left(\frac{R_c}{R_m}\right)=0.9\\&\left(2.9\times10^4\le\left(ME_0^{-1.5}R_e^3\right)<2.0\times10^6\right)\\&(1.0\,\text{mm}\le d\le3.0\,\text{mm})\end{aligned}\tag{3.29}$$

The comparison between the equations and data are shown in Fig. 3.36. In the case of turbine oil, the critical radius is largely affected by Eötvös number in comparison with that of water.

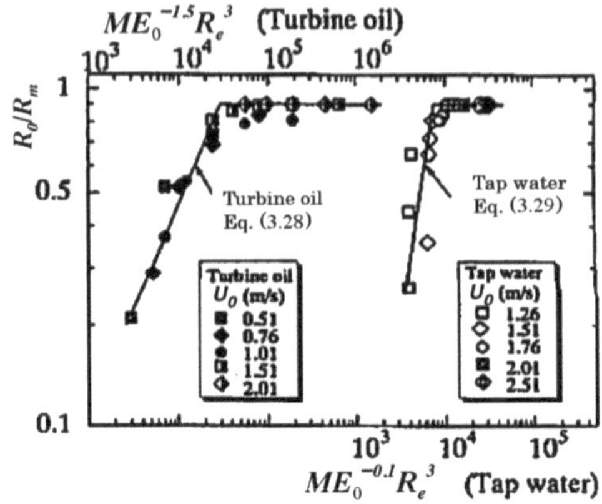

Figure 3.36 ***Comparison between experimental results and calculation results.***

In general, for practical application, it may be an important issue to predict the concentration radius of bubble, which is the distance from the central axis to the point of focus. Although the calculation result for the convergence of the bubble exists two patterns, it is targeted the pattern keeping one's focus narrowing.

Fig. 3.37 shows the relationship between the convergence radius and the bubble diameter. The dotted lines in the figure are the result for turbine oil. It shows a reasonable result that the focus radius increases with the increase of mean velocity of liquid or diameter of bubble. That is, a air bubble is concentrated to the central axis. The result is rearranged using nondimensional parameter, as shown in Fig. 3.38. The solid lines in the figure are given by the equations

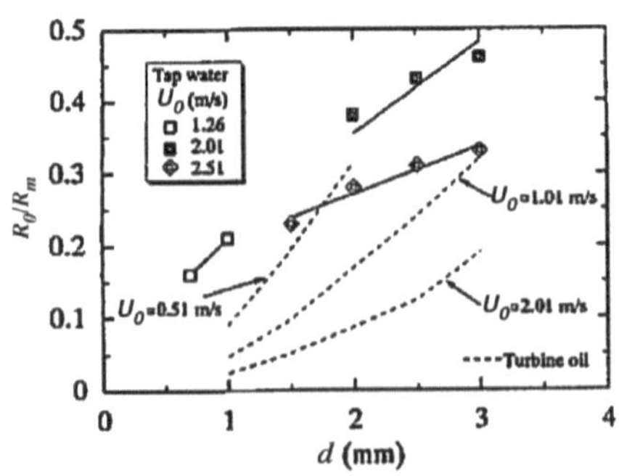

Figure 3.37 ***Convergence radius of air bubbles in tap water***

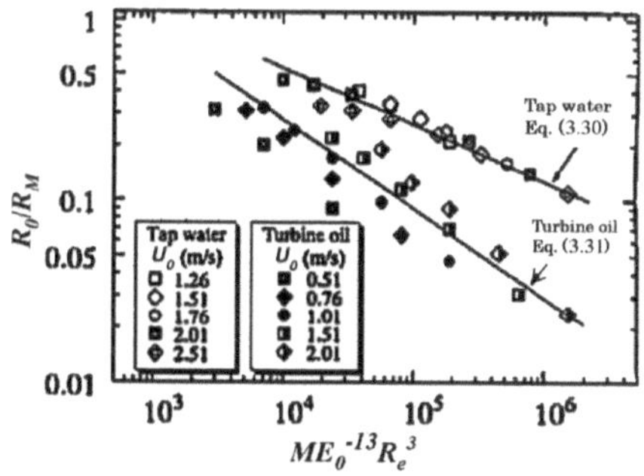

Figure 3.38 ***Verification of Eq. (3.30) and Eq. (3.31).***

For bubble in water;

$$\left(\frac{R_t}{R_m}\right) = 9.55 \times \left(M^{-0.31} E_0^{0.47} R_e^{-0.93}\right)$$
$$\left(7.0 \times 10^3 \leq \left(M E_0^{-1.5} R_e^3\right) < 2.0 \times 10^6\right) \tag{3.30}$$
$$(0.7\,\text{mm} \leq d \leq 3.0\,\text{mm})$$

For turbine oil;

$$\left(\frac{R_t}{R_m}\right) = 1.46 \times \left(M^{-0.13} E_0^{0.20} R_e^{-0.39}\right)$$
$$\left(3.0 \times 10^3 \leq \left(M E_0^{-1.5} R_e^3\right) < 2.0 \times 10^6\right) \tag{3.31}$$
$$(1.0\,\text{mm} \leq d \leq 3.0\,\text{mm})$$

Although the flow characteristics of gas–liquid two-phase flow has been reported on a pipe with spiral rib by Zarnett and Charles[26] there are few experimental studies about air bubbles in a deformed pipe. This calculated result hint at the possibility of the application to a bubble separation.

The drag reduction phenomenon in a heterogeneous flow is mainly given a description of the characteristics of liquid–solid, gas–solid and liquid–gas two phase flow in spiral pipe. However, many difficulties remain in the terrain to clarify the phenomenon yet because there are many number of factor affecting to the drag of flow. Down the road, it is expected that the complex phenomenon is raveled by a application of a spiral pipe.

REFERENCES

1. Charles ME, Cheh CH-SL, Chu H-L. *Can J Chem Eng* 1971;**49**(6):737.
2. Watanabe K, et al. *Bull JSME* 1984;**27**(228):1105.
3. Watanabe K, et al. *Bull JSME* 1984;**27**(230):1682.
4. Singh VP, Charles ME. *Can J Chem Eng* 1976;**54**:249.
5. Watanabe K, Uchida N, Kato H. *Trans JSME (in Japanese)* 1988;**54**(508):3362.
6. Ito H. *Trans JSME (in Japanese)* 1951;**17**(60):99.
7. Shimpton RJ, Granze HR. *Proc. Am. Soc. Civ. Eng., J. Hydraulics Div.* 1976;**102**(9):1343.
8. Ito H. *Trans ASME J Basic Eng* 1960;**82**(1):131.
9. Humphrey JAC, et al. *J Fluid Mech* 1981;**103**:443.
10. Taylor AKMP, et al. *Trans ASME J Fluid Eng* 1982;**104**:350.
11. Watanabe K, Hirakawa M, Kato H. *Trans JSME (in Japanese)* 1991;**57**(538):1995.
12. Watanabe K, Kamoshida T, Kato H. *Experimental Heat Transfer, Fluid Mechanics and Thermodynamics*. Elsevier Science Pub. Co., Inc.; 1988 p.693.
13. Watanabe K, Itagaki Y, Kato H. *Trans JSME (in Japanese)* 1989;**55**(509):p.25.
14. Watanabe K, Uchida N, Kato H. *Trans JSME (in Japanese)* 1990;**56**(532):3740.
15. R. Pfeffer and R. S. Kane, Proc. of Inter. Conf. on Drag reduction, Paper No. F1 (1974), F1-1.
16. Rossetti SJ, Pfeffer R. *AIChE J* 1972;**18**:31.
17. R. S. Kane and R. Pfeffer, NASA Contractor Report, NASA CR-2267, (1973),1-87.
18. Soo SL, Trezek GJ. *Ind Eng Chem Fund* 1966;**5**:388.
19. Clark RH, et al. *Trans Inst Chem Eng* 1952;**30**:209.
20. Richardson JF, McLeman M. *Trans Inst Chem Eng* 1960;**38**:257.
21. Boyce HP, Blick EF. *Trans ASME J Basic Eng* 1970;**92**:495.
22. Coughran MT. *Trans ASME J Fluids Eng* 1988;**110**:222.
23. Watanabe K, Zhang J. *Trans JSME B (in Japanese)* 1998;**64**(627):3597.
24. Choen YC. *An introduction to computational fluid mechanics*. John Wiley and Sons; 1979.
25. Tomiyama A. *Trans JSME B (in Japanese)* 1990;**56**(532):3740.
26. Zarnett GD, Charles ME. *Can J Chem Eng* 1996;**47**:238–41.

CHAPTER 4

Drag Reduction in a Pseudo-Homogeneous Flow

Outline

4.1 Introduction 75
4.2 Biofiber Suspensions 76
4.3 Liquid–Solid Particle Suspensions 81
4.3.1 Flow in a Pipe 82
4.3.2 Flow in a Divergent Channel 94
4.3.3 Flow over a Cylinder and Sphere 101
4.4 Measurement for Lift Acting on a Sphere 114
References 121

4.1 INTRODUCTION

When the size of a solid included in a liquid is fine, the flow behavior can be approximated as that of a homogeneous flow. For example, the diameter of a colloidal particle is 1–100 nm in the condition of being a colloid. From some previous studies, it is noticed that a certain kind of fibers or fine solid particles behaves as drag reducing additives in turbulent flow. Since it is 1–10 μm in order of the characteristics length, the sedimentation does not suddenly occur in liquid and the material has satisfactory follow ability to a flow. These mixtures are called slurry or suspension and they are not called solution. The flow is analyzed frequently by considering it to be a pseudohomogeneous fluid.

In general, the mixtures occur a nonlinearity in the viscous behavior with an increase of the concentrations of mixed matters. The viscous behavior is approximated by non-Newtonian fluid rheological equation, and the drag for a flow is deduced using a momentum equation under the boundary conditions. However, some experimental results of the suspensions in low concentration range show that drag reduction phenomenon occurs in the external flow with the separation. In this case, we may notice that the role of a particle to the flow behavior cannot not ignore.

In this chapter, the drag reduction of biofiber suspensions in a pipe flow is described at first, as it has a merit in the characteristics of degradation and the environmental loading for utilizing hydraulic transportation pipeline systems. Also, the flow behavior of fine particle suspensions

Drag Reduction of Complex Mixtures
http://dx.doi.org/10.1016/B978-0-12-809920-9.00004-4

is experimentally shown for flow in a pipe, flow around a sphere and flow around a cylinder, etc.

4.2 BIOFIBER SUSPENSIONS

It is known that drag reduction occurs in some types of fiber suspensions.[1] These fibers seem to be several micrometers in diameter and the size of the drag reducing additives is larger than that of high molecular weight polymers. Mih and Parker[2] measured the velocity profile of rayon and hardwood-kraft suspensions in a pipe using the annular purge impact tube. They discussed a phenomenological concept of turbulent flow for fiber suspensions in pipes suggested by the velocity profile data. Then, Kato and Mizunuma[3] analyzed the friction factor of artificial fiber and asbestos suspensions using measured viscosity data. Although the analogous expression of the friction factor of fiber suspensions based on the relationship between the velocity profile and the pressure loss in a pipe is given by these studies, the experimental data on drag reduction are lacking for other types of fiber suspensions. And it is known that polymer and surfactant solutions have a week point for the characteristics of degradation and environmental loading, respectively. Thus, it is an important to grasp the trend of the friction data of biofiber suspensions from viewpoint of a less load on environment

We examined experimentally the drag reduction phenomenon of four types of biofibers.[4] Fig. 4.1 shows the microphotographs of the fiber. The shape of each fiber is linear and the thinnest fiber is cotton yarn. Table 4.1 lists the physical constants of these fibers. In Table 4.1 (l/d) is given by measurement of sampled fiber in Fig. 4.1. Young's modulus is reference sited.[5]

Pressure losses were measured in 15 mm diameter pipe, and it is 300 mm in the test section. The pressure losses of the suspensions were measured using a differential pressure transducer with an uncertainty of less than 1.5%. Fig. 4.2 shows the wall shear stress plotted against the mean velocity for water and the fiber suspensions, respectively. τ_w was determined from the pressure loss.

In Fig. 4.2, the solid lines denote the empirical relationship for water in the turbulent flow region, which is that the shear stress is proportional to the power of 1.75 for the mean velocity. This is known as the Blasius approximation. The shear stress of the bio-fiber suspensions shows decreased deviation from the water solvent, in order of silk threads, cotton yarn and hemp fiber suspensions. The dotted lines are plotted from the curved lines with a power of 1.75 for the mean velocity. The observations suggest that

Figure 4.1 ***Microphotographs of bio-fibers.*** (A) Silk-A, (B) Silk-B, (C) Hemp, (D) Cotton.

Table 4.1 Dimensions of bio-fibers

Fiber	**Mean diameter** ***d*** **(μm)**	**Aspect ratio** ***l/d*****(−)**	**Young's modulus** ***E*** **(GPa)**
Silk-A	64.3	13.1	11.76
Silk-B	89.2	2.2	11.76
Hemp	32.5	26.3	35.28
Cotton	15.8	52.0	9.80

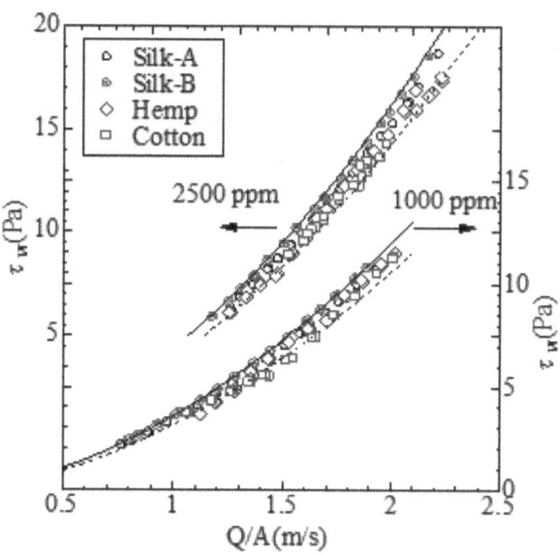

Figure 4.2 ***Shear stresses.***

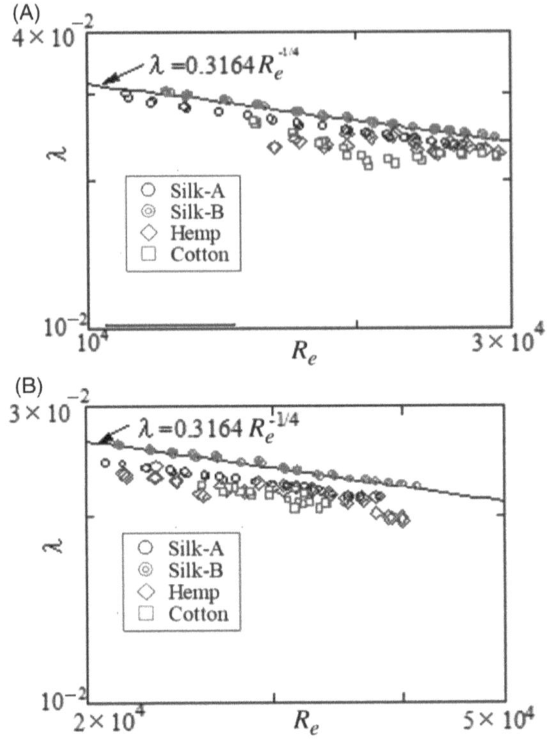

Figure 4.3 ***Friction factors.*** (A) $C_v = 1000$ ppm. (B) $C_v = 2500$ ppm.

the behavior of suspensions on the friction in turbulent flow regions is analogous to that of water versus Reynolds number.

Fig. 4.3A and B show the friction factor λ plotted against the Reynolds number Re for these fiber suspensions. These data exhibit characteristics typical of Type B[6] drag reduction, where the friction factor variation as the Reynolds number increases is almost parallel to that of Newtonian flow. The drag reduction values quoted in Fig. 4.3A and B at several Reynolds numbers were calculated based on the friction factor of the fiber suspensions and the friction factor for water at the same Reynolds number based on the Blasius. That is, *DR* is $DR = \{(\lambda_n - \lambda_s) / \lambda_n\} \times 100(\%)$ that the subscripts *n* and *s* refer the Newtonian and fiber suspensions, respectively.

The drag reduction ratio plotted against the concentration C_w and the aspect ratio l/d is shown in Figs. 4.4 and 4.5 respectively. In Fig. 4.4, the experimental data show that a peak in the drag reduction ratio exists, although it increases with the concentration of the suspensions. Because the

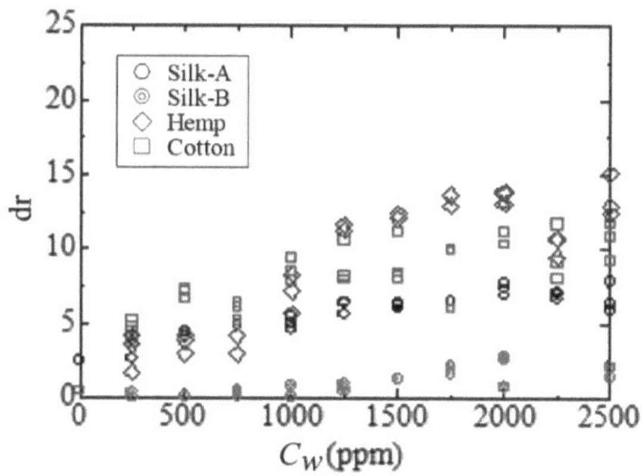

Figure 4.4 ***Drag reduction ratio versus the concentration.***

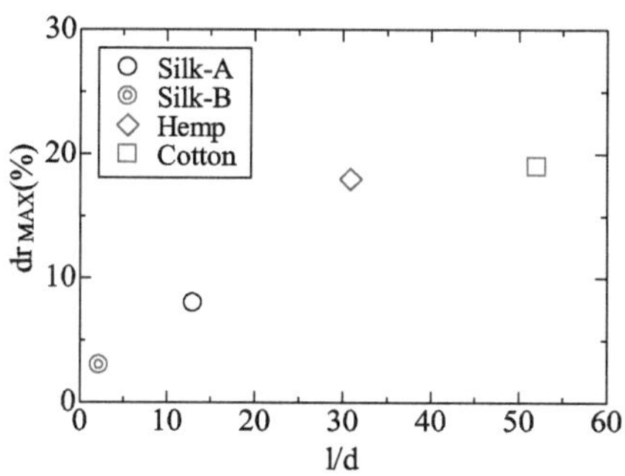

Figure 4.5 ***Drag reduction ratio versus* l/d.**

drag reduction occurs in the turbulent flow range, qualitatively, it can be considered that the dominant factor is turbulence modification by the fiber. Experimental data demonstrated that drag reduction increases in order of silken threads, cotton yarn and hemp fibers.Young's modulus will be one of the most important factors in the physical constants to distinguish the difference on the type of the fiber.

In every flow problem, there is some relationship between the drag reduction ratio and the concentration is involved and frequently the effect of (l/d) is also an important.To calculate the value of the reduction ratio, it may be assumed that an equation for the drag reduction ratio exists for any bio-fiber suspension under consideration

$$\left(\frac{\lambda_n-\lambda_s}{\lambda_n}\right)=\phi\left(C_w,\frac{l}{d},\frac{E}{\rho U^2/2}\right) \tag{4.1}$$

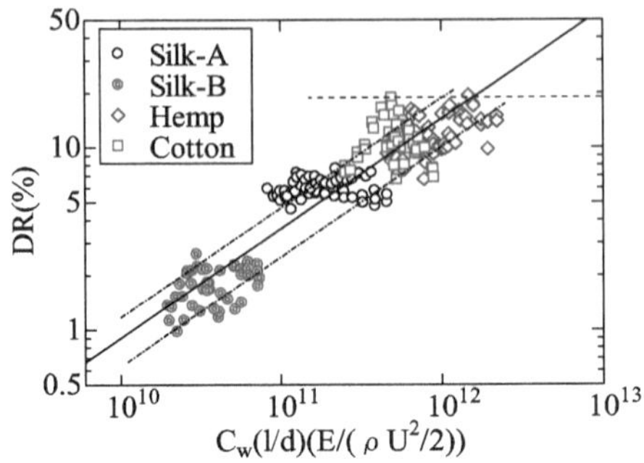

Figure 4.6 ***Correlation of drag reduction ratio.***

Fig. 4.6 shows the result, which was rearranged using the value is obtained for the product of these parameters. The solid line in Fig. 4.6 comes from the following equation,

$$DR(\%) = 9.0\times10^{-7}\left[C_w \times\left(\frac{l}{d}\right)\times\left(\frac{E}{\rho U^2/2}\right)\right]^{0.6} \tag{4.2}$$

In Fig. 4.6, the dotted lines show the error span ±30%. A good correlation is shown between the experimental data and the parameters in the range. Although difficulty remains in the arrangement of the experimental drag reduction data of the suspensions, it can be considered that the parameter which depend on the type of the biofiber.

On the other hand, the fiber suspensions have the additional advantage of no degradation for the mechanical shear by a pump circulating system. Fig. 4.7 shows the experimental results of the degradation test. Consequently, the degradation does not occur in these suspensions. By the experimental pipeline systems with continuously operating pump, no degradation of the suspensions is confirmed with even 8 hours as expected.

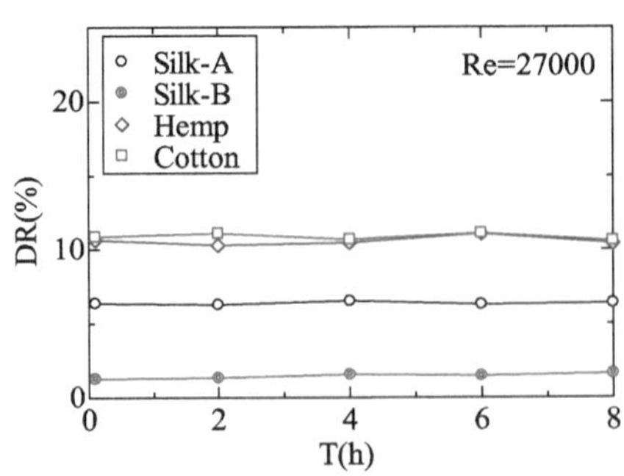

Figure 4.7 ***Drag reduction ratio versus the pump running hour.***

Figure 4.8 ***Micrographs of fine solid particles.*** (A) Carbon black particles. (B) Kaolin particles. (C) SiO_2 particles.

4.3 LIQUID–SOLID PARTICLE SUSPENSIONS

In case the shape of particle in fluids is spherical, it may be a typical suspension. Fig. 4.8A and B show the microphotograph of the carbon black, SiO_2, and the kaolin particles. It is seen that the shape of the carbon black particles is almost spherical and a particle chains it. The physical properties of these particles are given in Table 4.2. The mean diameter of carbon black, SiO_2, and kaolin equal the average value of about 100 particles taken from the microphotograph of the particle. The carbon black and SiO_2 particles are very similar for the physical properties in Table 4.2.

The flow curves[7] for carbon black and kaolin suspensions are shown in Fig. 4.9A–C obtained using laminar flow data in the pipeline system.

As described in Chapter 2, the power-law fluid model is the simplest available equation for relating shear stress to shear rate in non-Newtonian systems. Thus, the physical properties of viscosity are given by the flow curve obtained experimentally. The non-Newtonian property in viscosity of carbon black suspension occurs at the weight fraction above 2%. From

Table 4.2 Physical properties of fine particles

Particle	Mean diameter d_e (μm)	Desity ρ_s (kg/m³) × 10^{-3}
Carbonblack	1.29	1.75
SiO2	1.40	1.95
Kaolin	4.65	2.74
Fly ash	22.0	2.19

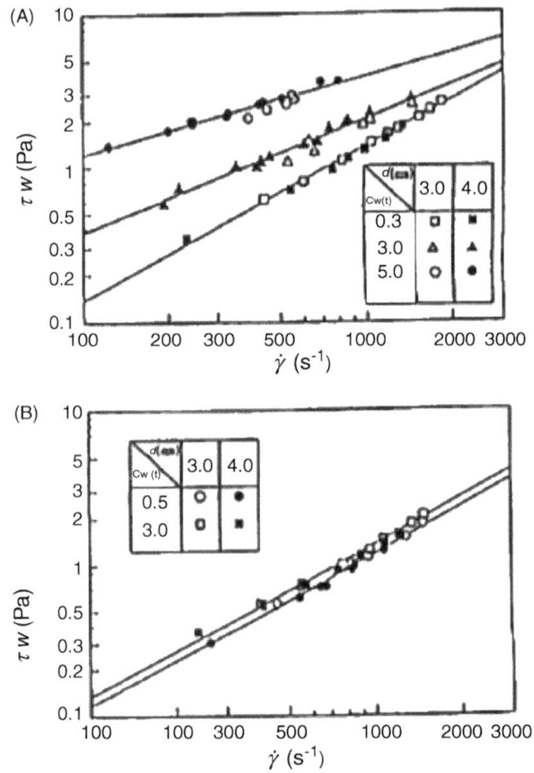

Figure 4.9 ***Flow curves of carbon black and kaolin suspensions.*** (A) Carbon black suspension. (B) Kaolin suspension.

Fig. 4.9A, we see that the experimental data conform with the equation for power law fluid in the range of $10^2 \leq \dot{\gamma} \leq 10^3$. The exponent in the power law, n', is readily determined as the slope of the solid line in Fig. 4.9A, and the values of n' is less than unity.

On the other hand, the rheological behavior for kaolin suspension is approximated as Newtonian fluid as shown in Fig. 4.9B, and the rheological behavior of the SiO_2 suspension is also approximated as Newtonian fluid. It is interesting phenomenon that the difference between rheological properties of carbon black and SiO_2 suspension occurs extremely in spite of the equivalent physical properties of the particle.

4.3.1 Flow in a Pipe

The experimental results of the friction factor of carbon black, kaolin, and SiO_2 suspensions for laminar and turbulent flow in a circular pipe, are shown in Fig. 4.10A–C. Dodge and Metzner[8] obtained the semitheoretical equation for friction factor of turbulent flow of power law fluid. It is given as

$$\frac{1}{\sqrt{f}} = \frac{4.0}{(n')^{0.75}} \log\left[R_e^{*} f^{1-n'/2}\right] - \frac{0.4}{n'^{1.2}} \tag{4.3}$$

where f is the friction factor and R_e^{*} is a generalized Reynolds number as defined as $R_e^{*} = \frac{D^{n'} U_m^{2-n'} \rho}{g_c K' 8^{n'-1}}$ n' and K' present the physical constants obtained using the flow curve described in Chapter 2.

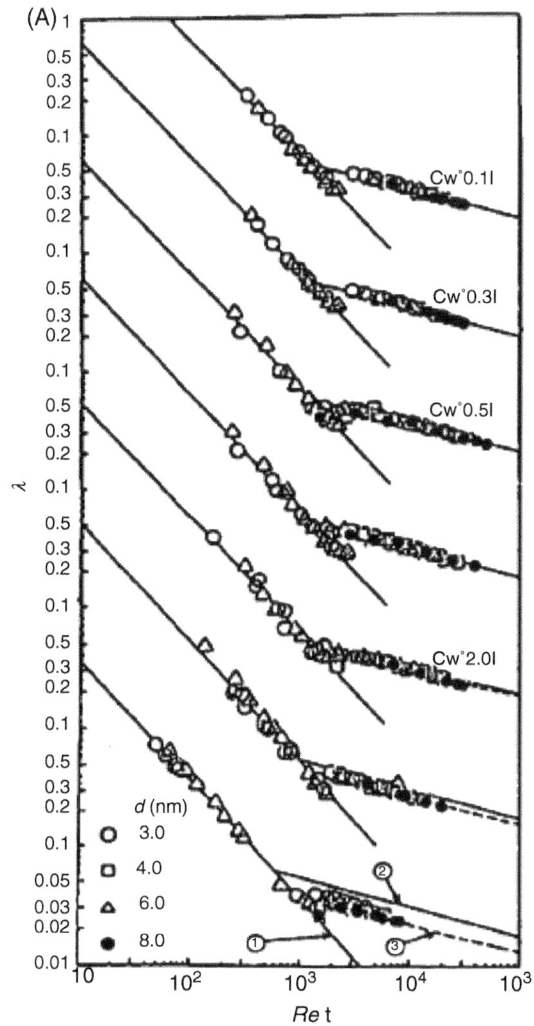

Figure 4.10 ***Friction factors.*** (A) Carbon black suspension.

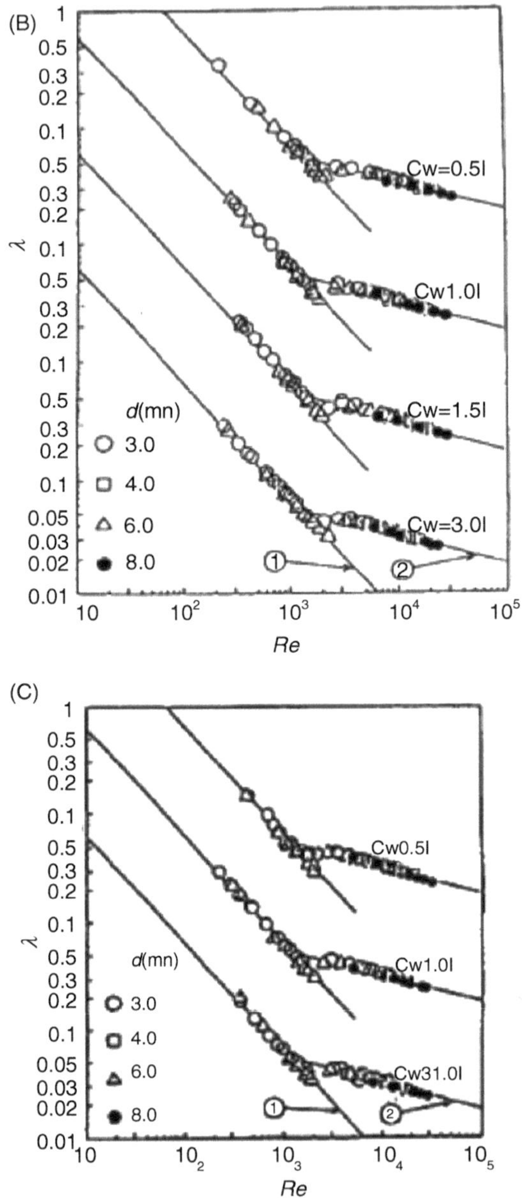

Figure 4.10 *cont.* (B) SiO_2 suspension. (C) Kaolin suspension.

If the friction factor that is defined by the pressure drop, λ is given as following equation by using the relationship of $\lambda = 4f$

$$\frac{1}{\sqrt{\lambda}} = \frac{2}{(n')^{0.75}} \log\left[R_e^* \left(\frac{1+3n'}{4n'} \right)^{-n'} 8^{1-n'} \left(\frac{\lambda}{4} \right)^{1-\frac{n'}{2}} \right] - \frac{0.2}{(n')^{1.2}} \tag{4.4}$$

where R_e^* is defined as $R_e^* = (\rho D^{n'} U_m^{2-n'} / \eta_{ck})$ and η_{ck} is the apparent viscosity given as $\eta_{ck} = K'(\dot{\gamma})^{n'-1}$.

On the other hand, the analytical equation of the friction factor for laminar flow as given as the following equation.

$$\lambda = \frac{\left(4\,(3n + 1/n)^n\, 2^{n+1}\right)}{R_e^*} \tag{4.5}$$

where n is the power of power-law model shown in Eq. (4.5).

In Fig. 4.10, solid line ① is Eq. (4.5). Solid line and dotted line are the Blasius formula and Dodge–Metzner formula shown as Eq. (4.4), respectively. It is seen from Fig. 4.10A that the experimental data of carbon black suspension for laminar and turbulent flow regions at the weight fraction under 1.0% agree with $\lambda = 64 / Re$ and Blasius formula for Newtonian fluid, respectively. However, the experimental data in the turbulent flow range at the weight fraction above 2.0%, whose behavior is expressed by power law fluid, decrease with increasing weight fraction compared with the Blasius formula. Flow behavior of carbon black suspension exhibits pseudoplastic and the friction factor agree well with Dodge–Metzner formula.

Experimental data for SiO_2 and kaolin suspension exhibiting near Newtonian behavior agree with solid line in laminar flow range and Blasius formula in turbulent flow range, respectively. Drag reduction phenomenon does not occur in either suspension of turbulent flow in a pipe. These results give that the drag reduction phenomenon may be explained by the characteristics of a psuedoplastic flow of the suspensions if we consider it from a viewpoint of the continuum approach. However, we cannot explain the reason of the drag reduction for the difference in the particle shape. The difficulty remains in the measurement suspensions for the relationship between the particle behavior and the turbulent structure because the carbon black suspension is a opaque liquid. If the particle has nanoscale microstructure, the suspensions may become a transparent liquid. In recent years, a nanoparticle of graphene oxide is expected to be applies in the field of various kinds

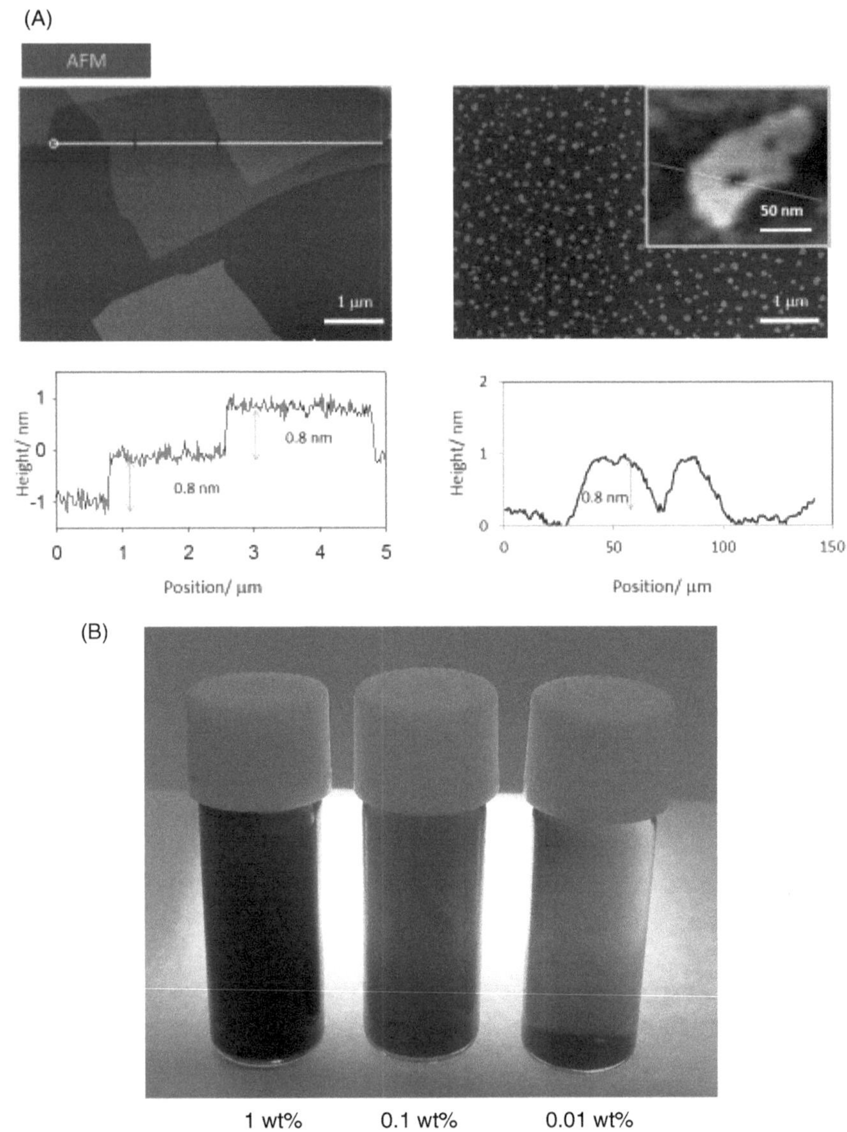

Figure 4.11 ***Graphene oxide particles and the suspensions.*** (A) Shape of graphene oxide particles. (B) State of the graphene oxide suspensions.

of industry because it is excellent in dispersion stability. However, there are few studies on the flow characteristics of the suspensions.

Fig. 4.11A and B show the tested nanoparticles of graphene oxide[9] and the suspensions. It is in the form of sheet which it is 8 nm in the thickness and 3 μm and 300 μm in the length, respectively, as shown in

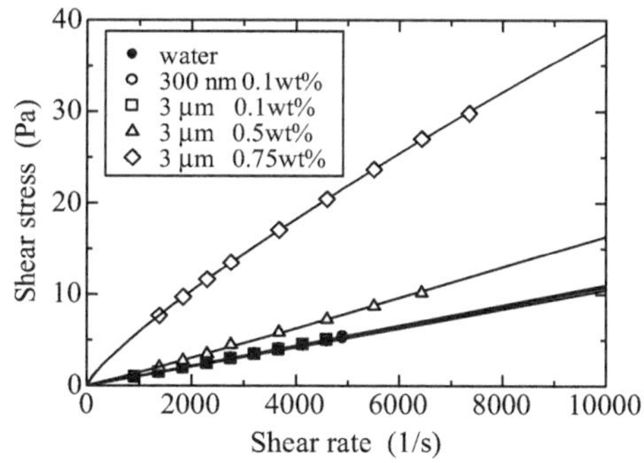

Figure 4.12 ***Flow curve of graphene oxide particle suspensions.***

Fig. 4.11A, and the transparency of the suspensions is kept in low concentration range as shown in Fig. 4.11B. As described above, the flow behavior of fine particle suspensions are capable of being treated as a psuedoplastic flow. Fig. 4.12 shows the flow curves of the suspensions. Experimental data are obtained by the pressure loss data, which were measured using a pipeline system with a 1.8 mm in pipe diameter. In the concentration range of 0.1%–0.5%, the linearity of the curve is satisfied and they are a Newtonian fluid. Non-Newtonian viscosity occurs in 0.75% suspensions.

The experimental result of the friction factors of suspensions are shown in Fig. 4.13. The figure is indicated by the rearranging method that is like to Fig. 4.10. R_e^* is a generalized a Reynolds number, and the solid line of laminar flow is given as Eq. (4.5). The dotted line in turbulent flow range indicates Eq. (4.4) that $n = 0.75$ is substituted for it. In the case of a 3 μm particle suspensions, the drag reduction phenomenon appears with an increase of the concentration. Fig. 4.14A and B indicate the relationships between drag reduction ratio and Reynolds number, and the concentration. Since the experimental data of suspensions shift to parallel with the Prandtl-Kármán formula for a Newtonian fluid friction, the drag reduction ratio is decreased by a certain percentage at turbulent range. The drag reduction ratio is slightly shown to be influenced by a pipe diameter.

It was experimentally clarified that the drag reduction occurs in suspensions of a bio-fiber, a carbon black particle and a graphene oxide nanoparticle at a certain concentration, however the influence of particle shape on the phenomenon is poorly understood. We need further study looking at the respect to the turbulent structure by measuring the velocity profile or the turbulence characteristics.

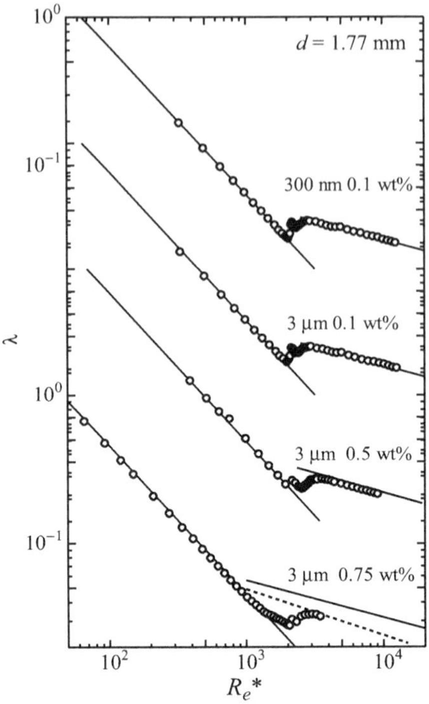

Figure 4.13 ***Friction factors.***

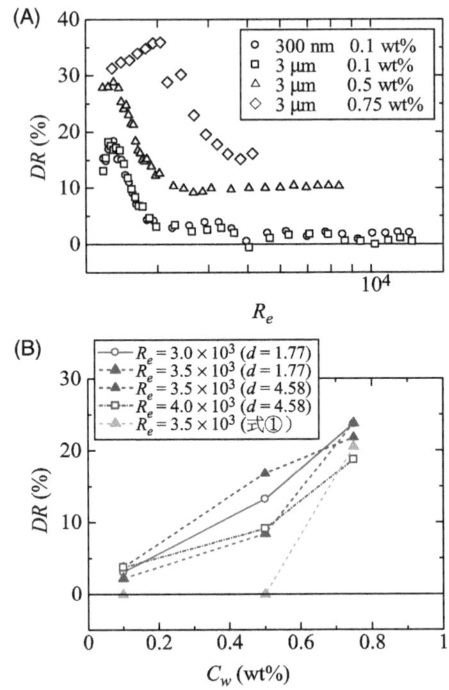

Figure 4.14 ***Drag reduction ratios.*** (A) Reduction ratio versus R_e. (B) Reduction ratio versus the concentrations.

The hydraulic transport[10,11] of fly ash slurries has been used to dispose of the fly ash at coal burning plants. Recently, fly ash slurries have also been used as a carrier fluid for coarse solid particles. An example is the hydraulic transport of fly ash slurries mixed with coarse solid particles for the refilling of a coal mine to prevent land erosion or fires. It was previously reported that pressure loss is reduced when a slurry mixture flows through a horizontal circular pipe.[12] It is important to grasp the phenomenon at drag reduction to economize energy during the hydraulic transport process. Furthermore, it is considered that the pressure loss can be further reduced by using a spiral tube[13] to transport the slurry. This could also prevent the sedimentation of solid particles from a mixture, but is very there has been little experimental evidence.

Fly ash suspensions behave as Bingham fluids within a range of concentration and especially at low shear rate. However, the power law equation may be modified for yield-pseudo-plastics at the low shear rate. Experimental data for fly ash suspension is given in Fig. 4.15. Rheological constants of power law fluid are obtained by the solid line in the figure. When the shear rate range for C_w = 30%–50% slurries is reduced and extrapolated, the flow curves exhibit the characteristics of power-law fluids in the range of $\dot{\gamma}_w = 10 - 10^2\,(S^{-1})$.

Although the friction factor of a spiral pipe for Newtonian fluids was obtained by Eq. (3.3) in Chapter 3, the friction factor of power-law fluids is derived from the similar method.

The velocity profile in the flow direction of an elliptic pipe,[14] V_3 is given as follows,

$$V_3 = \frac{n}{n+1}\left[\frac{\pi}{\eta_{ck} H}\left(-\frac{\partial p}{\partial x_s}\right)\right]^{\frac{1}{n}}\left\{1-\left(\frac{x_1^2}{a^2}+\frac{x_2^2}{b^2}\right)^{\frac{1+n}{2n}}\right\} \quad (4.6)$$

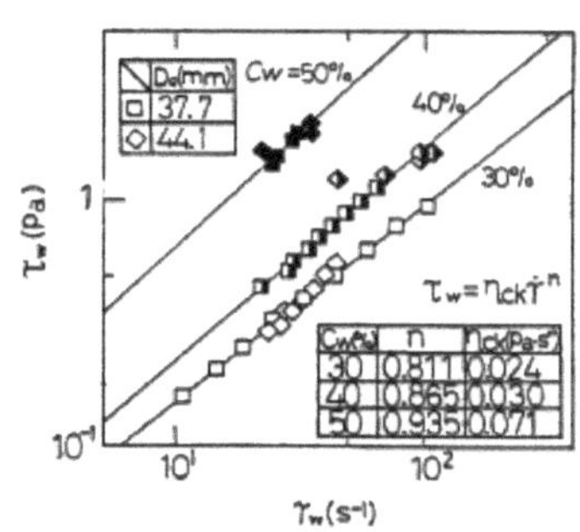

Figure 4.15 ***Flow curve of fly ash slurries.***

where x_3 is the pipe axis. x_1 and x_2 are the semiminor and the major axis, and H in Eq. (4.6) is

$$\mathrm{H} = \frac{\pi}{b^{n+1}}\left[2 + \frac{n+1}{2}(\beta^2 - 1) + \frac{3}{32}(n+1)\right.$$
$$\left.(\beta^2 - 1)^2 + \frac{1}{77}(n+1)(n-3)(\beta^2 - 1)^2 + \ldots\right] \tag{4.7}$$

where β equals (b/a) that is the ratio of the semiminor axis to the semimajor axis. By using the analytical technique described in Chapter 3, the relationship between the flow rate Q and the pressure gradient $(-\partial p/\partial x_3)$ for a straight pipe equal length is derived by integrating Eq. (4.6) over one half of the shadow area shown in Fig. 3.3.

The mean velocity in the straight pipe with the same spiral pipe section becomes

$$V_m = \frac{Q}{A} = \frac{n}{3n+1}\left[\frac{\pi}{\eta_{ck} H}\left(-\frac{\partial p}{\partial x_3}\right)\right]^{\frac{1}{n}} \tag{4.8}$$

From Eq. (4.8), the friction factor, λ' is derived as

$$\lambda' = \frac{2De}{\rho f \rho}\left(\frac{3n+1}{n}\right)\frac{\eta_{ck}\mathrm{H}}{\pi} V_m^{n-2} \tag{4.9}$$

For the friction factor of the tested spiral pipe λ relates to λ'

$$\lambda = C_1 / R_e^{\ *} \tag{4.10}$$

where C_1 is

$$C_1 = \frac{2}{\cos\beta_s}\left(\frac{3n+1}{n}\right)\frac{HD_e^{n+1}}{\pi} \tag{4.11}$$

Thus, it is given as follows,

$$\lambda = \frac{2}{\cos\beta_s}\left(\frac{3n+1}{n}\right)\frac{HD_e^{n+1}}{\pi} \Big/ R_e^{\ *} \tag{4.12}$$

where β_s, D_e and n are the spiral angle, the equivalent diameter and the power law exponent, respectively. $R_e^{\ *}$ is a generalized Reynolds number defined as $R_e^{\ *}(\rho_f D_e^n V_m^{2-n} / \eta_{ck})$.

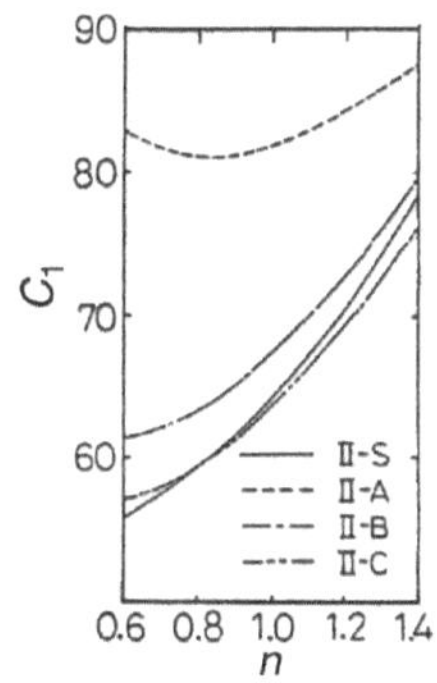

Figure 4.16 ***C_1 in Eq. (4.11) for a II-type spiral pipe.***

Eq. (4.12) deduces $\lambda = 64/R_e$ for Newtonian fluid in a circular straight pipe where $n = 1$ and $\beta_s = 0$. The relation between the factor, C_1 from Eq. (4.11) and the exponent in power-law fluids for a II-type spiral pipe is shown in Fig. 4.16 the figure suggests C_1 increase with the power, n except in the region of $n < 0.9$ for a A-type spiral pipe.

The comparison of Eq. (4.12) with the experimental data based on the rheological constants in Fig. 4.6 are shown in Fig. 4.17A–C. These figures show that Eq. (4.12) agrees with the experimental data at$10^2 \leq \mathrm{Re} \leq 10^3$. However, the data show considerable scatter and the correlation is highly sensitive to R_e^* in certain ranges. The scatter is probably due to the use of a characteristics constant of power law fluid. The analytical result based on the rheological equation for power-law fluid is restricted in its application, but is of interest because of its simplicity and qualitative significance.

Fly ash slurries have been used as carrier fluids for the coarse solid particles in hydraulic transport, and the drag reduction of the spiral pipe may also be expected from the point of view of economizing energy in the pipeline systems as shown in Fig. 3.23. As described in Chapter 3, it has been shown that the maximum drag reduction ratio obtained for a B-type spiral pipe at the critical pressure gradient was approximately 40% compared with that of circular pipe that carry water. There is no fixed theory which compass the flow, drag reduction and pressure loss in the pipe of slurries containing coarse solid particles mixed with fly ash to that of water at this time. It is thought that inference is caused by solid particles interacting with the turbulence of the mixture flow or by the pipe wall shear stresses of solid particles or by the characteristics of non-Newtonian fluids for the carrier fluids. Therefore, it is an important to clearly understand the flow separation moreover the drag of a particle in these suspensions.

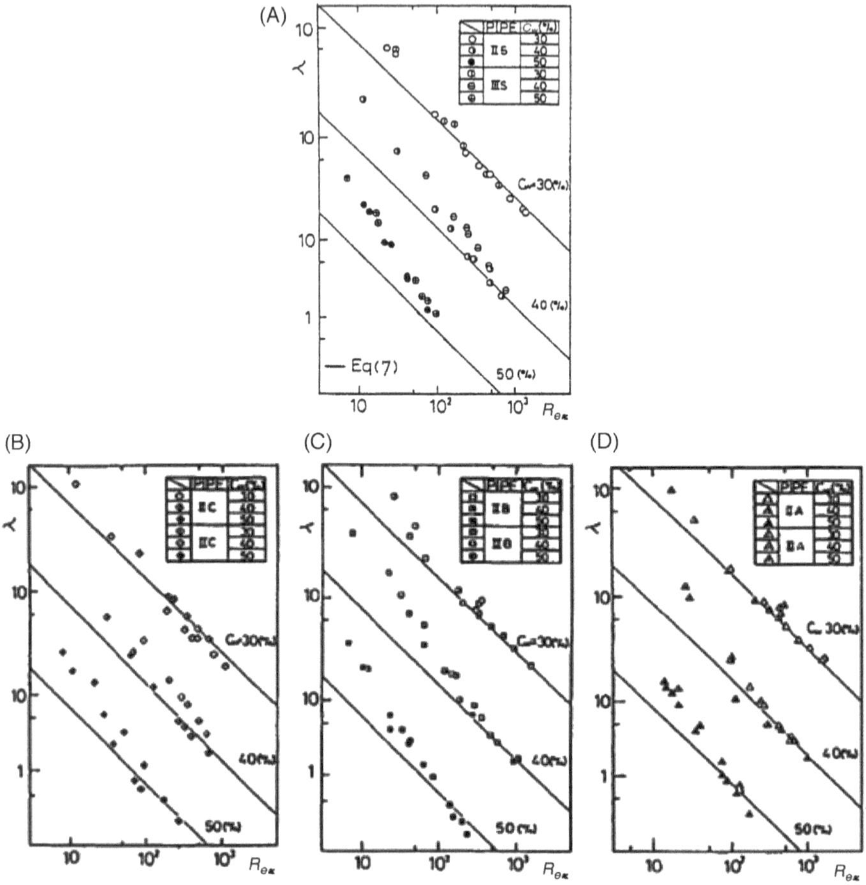

Figure 4.17 ***Friction factors of fly ash slurry.*** (A) Circular pipe. (B) Type-C pipe. (C) Type-B pipe. (D) Type-A pipe.

We now turn to a treatment of the circular pipe flow of fine solid particle suspension. It is a basic problem in fluid mechanics how does the turbulent transition from laminar to turbulent flows in a circular pipe. Experimental results, which related to criterion for the end of stable laminar flow, may give suggestion for the effect of fine particles on the stability of flow. The stability of laminar flow depends upon the flow conditions upstream of the pipe entrance. Experimental data on the friction factor were obtained using a circular pipe with bell mouth as shown in Fig. 4.18.

Fig. 4.19A and B shows the experimental results of the friction factor of carbon black and silica dioxide suspensions in the pipe with a bell-mouthed inlet and no bell-mouthed type. From Fig. 4.19A and B, it is seen that the

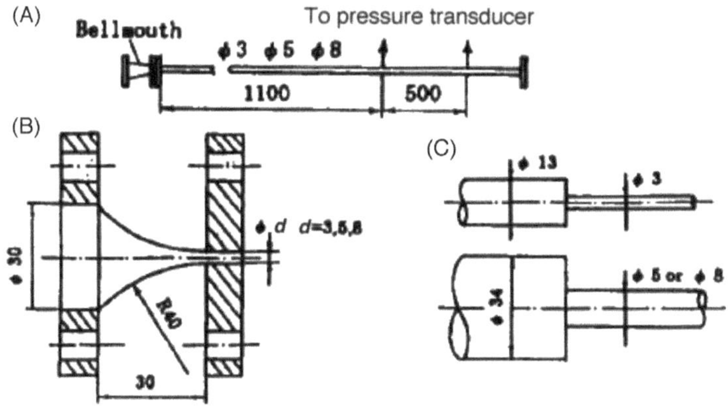

Figure 4.18 ***Shapes of pipe entrance.*** (A) Experimental pipeline system. (B) Shape of bell mouth. (C) Entrance condition without bell mouth.

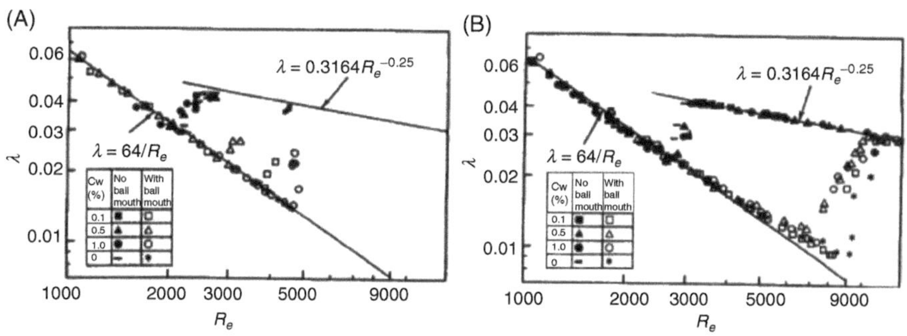

Figure 4.19 ***Friction factor for suspension in a pipe with bell mouth.*** (A) Carbon black suspensions (ϕ3 mm). (B) Silica dioxide suspensions (ϕ8 mm).

lower critical Reynolds number is almost equal to that of tap water or fine solid suspension. The value falls within the purview of R_{elc} = 2350–2700. For the stability of laminar whose upper critical Reynolds number increases, there is significant deviation as shown in Fig. 4.19A. The upper critical Reynolds of tap water number increase to Re_{ehc} = 4500 in D = 3 mm pipe with a bell-mouthed inlet, and it is Re_{ehc} = 9000 in D = 8 mm.

It is interesting to note that the upper critical Reynolds number of fine solid particle suspension decreases comparing with that of tap water and the concentration to minimize exists. As shown in Fig. 4.19A, the upper critical Reynolds number, R_{ehc} for water is$R_{ehc} \cong 4500$ For the case of carbon suspensions in 3 mm diameter pipe, the value for weight fraction of 0.1%, 0.5%, and 1.0% are 3800, 3000, and 4600, respectively. For the pipe of 8 mm in diameter, the critical Reynolds number is almost equal to 2800;

the upper critical Reynolds numbers for weight fraction of 0.1%, 0.5%, and 1.0% are 9000, 7600, and 7400, respectively. In contrast, the value decreases with the increase of the concentration. The results of silica dioxide particle suspensions show that for the pipe of both 5 mm and 8 mm in diameter, the upper critical Reynolds numbers for weight fraction of 0.1%, 0.5%, and 1.0% are 8300, 7400, and 7200, respectively.

The carbon black suspensions exhibit the non-Newtonian behavior in viscosity, however the silica dioxide suspensions don't show. The stability of turbulence is considered to be influenced by the characteristics such as the particle shape and size.

4.3.2 Flow in a Divergent Channel

The flow pattern in a divergent channel gives an important criterion in separating flow systems. Thus, the empirical flow visualization results are available to clarify the effect of fine particles on the flow separation. But, a direct examination of the separation zone was difficult because the fluids were opaque. To understand the flow behavior, the flow[15] was visualized in a two-dimensional diffuser, using computed sonography. To ensure flow separation, each had a half angle, α greater than 3.5 degrees specifically, the angle were 5 and 11 degrees. To generate planer motion for visualizing the flow, the diffuser was a truncated half wedge. It was made of clear plastic and had a half angle of 22 degrees. A computed sonography system was attached to the nonoblique wall of the half wedge, as shown in Fig. 4.20. Since the test fluids already contained particles, tracers were not needed. In this technique ultrasonic waves are sent into the flow by a sonographic probe and reflected by the solid particles. The reflected echo signals are processed by a computer system to create an image of the flow on the screen.

Flow in the half-wedge diffuser was designed to generate Reynolds numbers ranging from 1000 to 2000. In this way, the flow was initially laminar and therefore motion in the separation zone was steady and easily

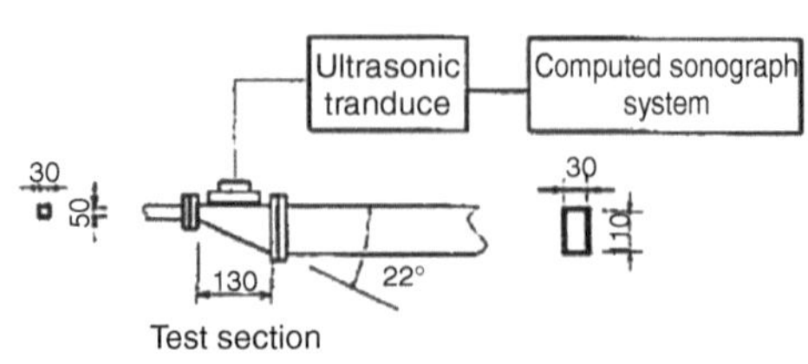

Figure 4.20 *Two dimension flow visualization system.*

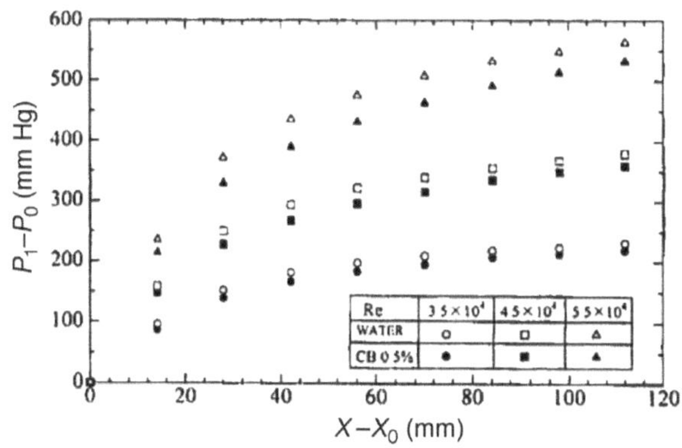

Figure 4.21 ***Comparison of the pressure profiles of tap water and the 0.5% carbon black suspension in the 5 degrees conical diffuser.***

observed. Reynolds number is defined as, $R_e = \rho V_m D/\mu$ where V_m is the average velocity at the inlet of a diffuser, D is the diameter (or equivalent diameter) of the diffuser inlet, and ρ *and* μ the density and viscosity of the test fluid. The velocity determined from in the flow rate, which was measured by collected over a known time. For the conical diffuser, the flow rate were much higher than those in the half-wedge diffuser in order to generate measurable pressure differences.

Fig. 4.21 presents direct measurement for a typical case, the 0.5wt% suspension in the 5 degrees diffuser. The ordinate is the pressure recovery, defined as P_i–P_0, where P_i is the pressure at the i th location in the diffuser (i= 1 – 8). P_0 is an upstream reference pressure. In the abscissa, X is the axial distance from the extended vertex of diffuser, and X_0 is the distance from the vertex to the entrance. Data are presented for three flow rates (in terms of Reynolds numbers), and the corresponding data for water are given as well. The graph shows that pressure recovery was less for the suspension than for water alone, and increasingly so as the flow rate increased. The reduced pressure recovery with particles suggests that the flow separated earlier in the diffuser. The same data are presented in dimensionless form in Fig. 4.22 along with data for the 11 degrees diffuser. The ordinate is the pressure recovery coefficient, which is the dimensionless quantity normally, used in such flows and which is defined as $C_p = (p_i - p_0)/\frac{1}{2}(\rho V_m^2)$. The abscissa is the area ratio $R = S_i/S_0$, where S_i is the cross-sectional area at the i th pressure tap and S_0 is the area at the inlet. R was chosen as the dimensionless axial distance because C_p can be expressed in terms of R for an ideal

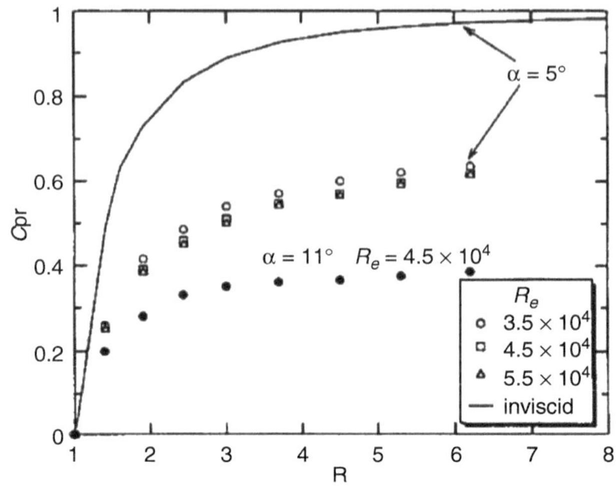

Figure 4.22 ***Pressure recovery coefficient for a carbon black suspension (C_w = 0.5%).***

(inviscid) flow, which does not separate, that is, $C_p = 1 - 1/R^2$. By expressing the results in terms of C_p the 5 degrees data in Fig. 4.22 are almost collapsed, which is expected because the Reynolds numbers are of order 10^4. Also shown in the graph are data for the wider-angle diffuser ($\alpha = 11$ degrees), as well as the curve for an ideal fluid. With the wider angle, it is expected that less pressure is recovered and the data confirm expectation.

Particle effects can be seen more clearly if differences between the results for water and the suspensions are presented. To this end, let ε designate the percentage change in the pressure recovery coefficient, defined as $\varepsilon = \{(C_p - C_{p0})/C_{p0}\} \times 100\%$, where C_{p0} is the pressure recovery coefficient for water. The pressure measurements are presented in this form to illustrate, in Fig. 4.23A the effect of Reynolds numbers and, in Fig. 4.23B the effect of particle concentration, for the two half angle respectively. The figure shows that the pressure coefficient can be reduced by as much as 13% and the reduction is more extensive in 11 degrees cone. The influence of Reynolds number, shown in Fig. 4.23A, is surprisingly strong when one considers that the particles are micron-sized and that the flow is dominated by inertia. The graph of Fig. 4.23B show that the largest changes in pressure recovery occur with the intermediate concentration, 0.5wt%, that is, the maximum effect occurs at a solid volume fraction of about 0.0025. This latter result depends the mystery as to how the particles affect the flow because, whatever the mechanics, one expects the hydrodynamic effect to increase with concentration in the dilute regime, in the same way that the viscosity increases

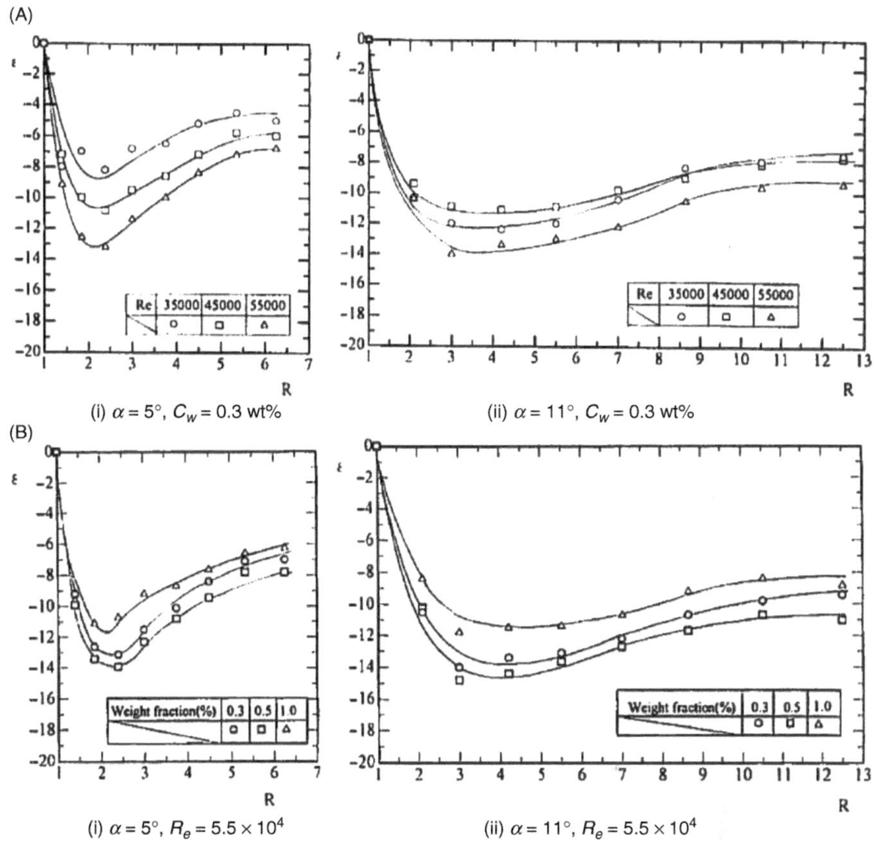

Figure 4.23 ***Reduction ratio for pressure recovery coefficient.*** (A) Effect of Reynolds number. (B) Effect of particle concentration.

in that regime. Indeed, because the viscosity is known to increase linearly with concentration for solid volume fraction up to about 0.1, one expects a similar increase for any other hydrodynamic effect. One does not expect the effect to structure when the solid volume fraction is only 0.0025. The behavior is consistent, however, with the prior results for sphere, the data in Watanabe et al.[7] show that drag reduction was maximum for a concentration between 0.5% and 1.0%, for carbon black particles.

Visualization using the computed sonography system confirmed that the particles caused early separation. Fig. 4.24A and B illustrate images obtained by the system for water and 0.1wt% carbon black suspension.

With water, tracer particles had to be added and these were 12-μm metallic coated particles at a weight concentration of 3.2×10^{-3} kg/m^{-3},

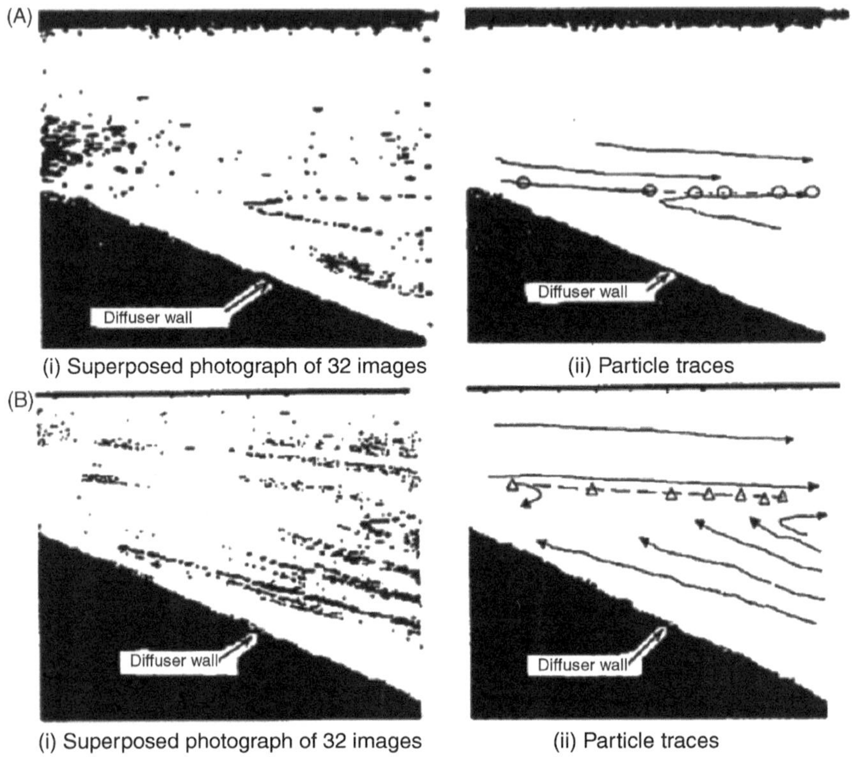

Figure 4.24 ***Flow visualization resultsfor aasymmetric diffuser.*** (A) Tap water. (B) Carbon black suspension.

which translates to weight concentration of 0.0003%. This value is so far below the minimum carbon black concentration of 0.1% that the tracer particles did not likely affect the flow of water. The first image in Fig. 4.24A and B is a photograph obtained by superposing at least 30 video pictures. The streaks in each photograph were used to construct the streamline pattern, which is the second image of each figure. In each second image, the dividing streamline is shown by a dashed line.

Points on the separation streamlines are plotted in Fig. 4.25, for several particle concentrations and for two Reynolds numbers. In these plots, X is the distance along the upper (nonoblique) wall and Y is the distance perpendicular to the wall. To show the separation streamline clearly, the perpendicular scale is magnified and hence the lower (oblique) wall appears much steeper than its actual 22 degree. As shown in Fig. 4.25, curves were fitted through the data points and extrapolated to the wall to identify the points of separation. The plots clearly show that the suspensions separated

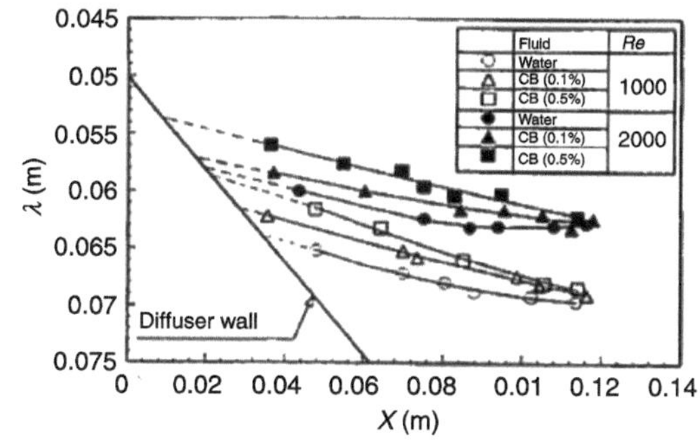

Figure 4.25 ***Separation location of the asymmetric diffuser.***

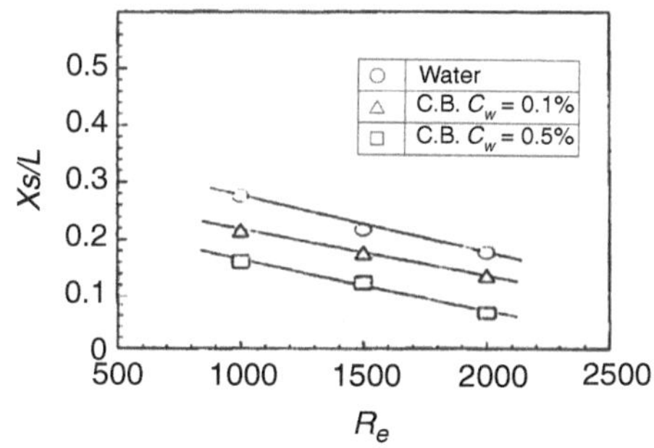

Figure 4.26 ***Dependence of the separation location on the Reynolds number.***

earlier than water, as anticipated. Furthermore, the 0.5% suspension separated earlier than the 0.1% suspension, consistent with the pressure recovery results. As expected, flows at the higher Reynolds number separated earlier, for all three fluids.

The separation results are summarized in Fig. 4.26 in which the distance to the separation point, X_s, is made dimensionless by the length of the diffuser channel, taken along the oblique wall. The graph reinforces how the separation point moves upstream with Reynolds number and with particle concentration. Fig. 4.27 shows values of ε for other diffuser, the half-wedge diffuser, for Reynolds number of 1000 and 2000. The results are similar to those for the conical diffuser, numerically they are somewhat larger, perhaps because the entering flow was laminar in the half-wedge case.

As described above, this experiment demonstrates that carbon black particle can cause early flow separation in a divergent channel and thus decrease the recovery of pressure. At this concentration level, it is remarkable that the particles had any effect at all. Since the particles did not change the viscosity in a significant way, one must consider other mechanisms.

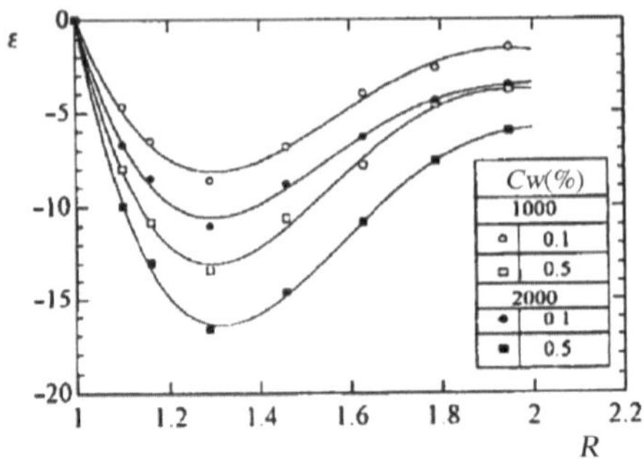

Figure 4.27 ***Reduction ratio for pressure recovery coefficient at*** **R_e** ***= 1000 and 2000.***

One possibility is that the particles had a roughness effect like surface. That is, when particles are attached to a surface to create roughness elements, they can influence flow separation and the transition to turbulence. Classic work by Fage and Warsap,[16] as reported in Schlichting,[17] shows that the drag on a circular cylinders is affected when the size of roughness elements is $O(10^{-3})$of the cylinder diameter. Their graph shows that when the Reynolds number is below about 3×10^5, that is, before transition to turbulence takes place in the boundary layer, roughness elements cause early transition and hence later separation and lower drag. Another classical work, which is relevant, is the study of roughness and transition in boundary layers flow on a flat plate. Data collected by Dryden, et al.,[18] show that transition to turbulence occurs early when the roughness size is order of the displacement thickness of the boundary layer. The size criterion is the same as that in the study of Fage and Warsap, where the relative roughness size was of order 10^{-3}. This is, the Reynolds number in their cylinder flows were of order 10^5 and, since the boundary layer thickness on a cylinder is of order, $Re^{1/2}$ the relative boundary layer thickness was of order $10^{-5/2}$ and therefore comparable with the relative roughness size of 10^{-3} Hence, classical work is consistent in showing that boundary layer flow is affected by surface roughness when the irregularities are size of the boundary layer thickness.

But, the particles in this experimental data were one micron in size and the thickness of the boundary layer in the diffuser was approximately 1/10 of the 13 mm inlet diameter, or about 1 mm. Hence the particles were two or three orders of magnitude below the size known to influence the flow. Furthermore, the particles in this flow were not fixed to the surface but rolled along it, and their effect would be even smaller on that account.

Another mechanism that may be related to the present findings is particle migration. Particle movement across streamlines and approach to a wall or away from a wall may be caused by particle interactions or by inertial effects. Interaction-caused migration has been an active research topic recently and a good reference is the seminal paper by Leighton and Acrivos.[19] This type of migration occurs with concentrated suspensions; in the referenced paper and in other such work, solid volume fractions are in the range of 0.1–0.5. The fractions in this, however, range from 0.0005 to 0.005, and so migration due to particle interaction is negligible in this data. For suspensions as dilute as this experimental result, the inertial-driven migration may be relevant. It is not yet possible to predict the effect of particles on the flow field but the experimental results shown in the following Section 4.4 may be developed both for the separation of a suspension flow and for the migration of particles.

4.3.3 Flow over a Cylinder and Sphere

When bodies are completely surrounded by a flowing fluid, the flows are called as a external flow. Such a body may be subject to kind of forces arising from relative motion between the body and the fluid. These forces are termed the drag and the lift. As for the drag force, it is given by addition of the friction drag and the pressure drag acting on the surface. Because the drag is affected by flow pattern, the drag reduction data may give also some considerations on the flow behavior. In general, it is very difficult to visualize the flow around a cylinder or a sphere because fine solid particle suspension is not transparent.

On the other hand, the pressure drag is dominant for total drag in high Reynolds number range and the separation point is shown by the experimental data of the pressure distribution.

The surface pressure field of flow around a circular cylinder is described by the pressure coefficient, $C_p = (p - p_\infty) \Big/ \frac{1}{2}\rho U_m^2$, where p is the pressure at surface of circular cylinder, p_∞ is free stream static pressure, ρ is fluid density and U_m is uniform velocity at the front of a cylinder.

Fig. 4.28 shows the effect of the concentration of carbon black suspensions on the pressure distribution[20] for $R_e = 3000$. As shown in Fig. 4.28, the separation angle increases with an increase of the concentration, however it decreases in the case of $C_w = 1.0\%$ and is almost equal to that of tap water. The experimental results for the separation angle that are obtained from the pressure distribution are illustrated in Fig. 4.29. Since the flow region little affected by shear stress for the drag, we can estimate the drag of circular cylinder by the pressure drag coefficient,

$$C_d = \int_0^\pi (p - p_\infty)\cos\theta \, d\theta \Big/ \frac{1}{2}\rho U_m^2 \tag{4.13}$$

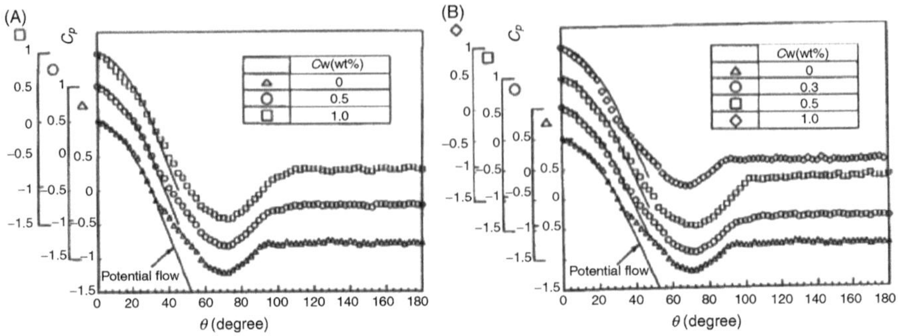

Figure 4.28 ***Pressure coefficient of a circular cylinder.*** (A) Silica dioxide suspensions. (B) Carbon black suspensions.

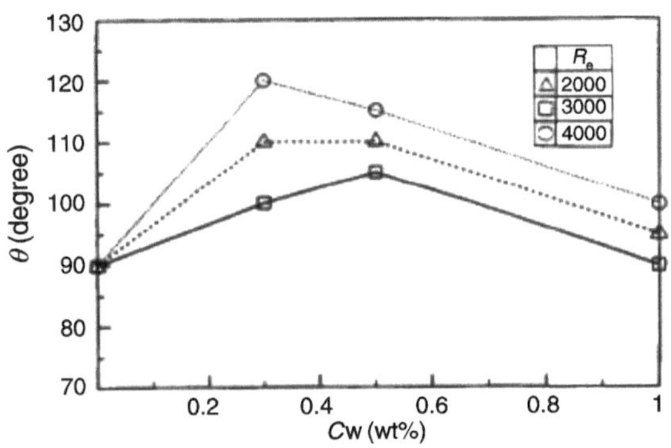

Figure 4.29 ***Separation angle for carbon black suspension.***

Drag reduction ratios ε for carbon black suspension defined as $\varepsilon = [(C_{dw} - C_{ds}) / C_{dw}] \times 100(\%)$ are plotted in Fig. 4.30, where C_{dw} and C_{ds} the pressure drag coefficient for tap water and the suspension, respectively. It is clear that the most effective concentration exists to the drag reduction phenomenon. Although the drag of a circular cylinder or a sphere is estimated indirectly by integrating the pressure distribution when the pressure drag is the major part of the total drag, we can measure directly the total drag by applying a free-fall body system.

Fig. 4.31 shows the measurement system.[21] It is consisted of a tank and a sensor for fall velocity measurement. The tested sphere is dropped without imparting an initial velocity or rolling of the sphere into the tank. The fall velocity is measured by using a ultrasonic transducer because the suspension is a nebulous liquid. Dimensions of a sphere are listed in Table 4.3.

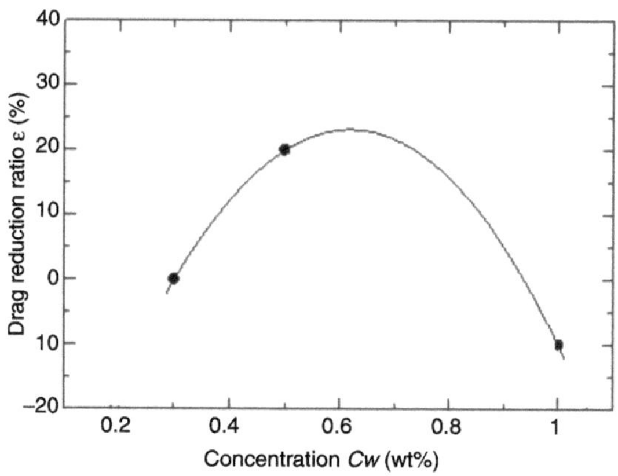

Figure 4.30 ***Drag reduction ratio for the pressure coefficient of carbon black suspension.***

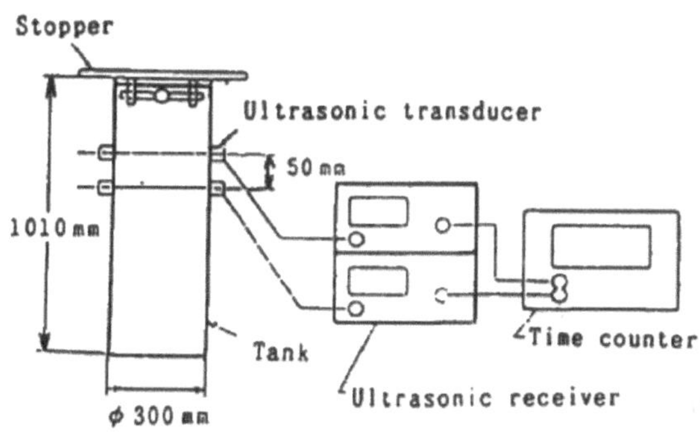

Figure 4.31 ***Experimental apparatus for falling sphere test.***

Table 4.3 Dimensions of Test sphere

No.	Material	Diameter d_s (mm)	Density ρ_s (kg/m^3) × 10^{-3}
S4	Acrylic resin	9.85	1.182
S7	Glass	9.53	2.518
S9	Stainless	7.94	7.671
S17	Brass	12.70	8.458
S19	Tungsten	7.93	14.947
S20	Tungsten	12.70	14.855

To clarify the drag reduction phenomenon in suspensions, at first, we must obtain the analytical value based on that in Newtonian fluid for the fall velocity of a sphere in the acceleration range because it is a standard value.

The equation of motion considering the Basset term for a single sphere in Newtonian fluid is

$$\begin{aligned}(m_s + km_f)\frac{dV_s}{dt} &= (m_s - m_f)g - C_d \pi \rho_f \frac{d_s^2}{8} \\ V_s^2 &- \frac{3}{2} d_s^2 \sqrt{\pi \rho_f \mu} \int_0^t \frac{(dV_s/dt)}{\sqrt{t-\tau}} d\tau\end{aligned} \tag{4.14}$$

where m_s and km_f are the mass and the apparent mass of the sphere, respectively. V_s is a fall velocity of a sphere. Because it is impossible to obtain an analytical solution of V_s, we can solve Eq. (4.14) numerically in the same manner as Asakura et al.[22] Eq. (4.14) contains the drag coefficient, C_d, which assuming a steady-state value, is a function of the Reynolds number;

Stokes

$$\begin{gathered}C_d = \frac{24}{R_e} \\ 0 \langle R_e \langle 1\end{gathered} \tag{4.15}$$

Schiller and Nauman[23]

$$\begin{gathered}C_d = \frac{24}{R_e}(1 + 0.15\, R_e^{0.687}) \\ 1 \le R_e \langle 700\end{gathered} \tag{4.16}$$

Carey[24]

$$\begin{gathered}C_d = \left[\left(\frac{24}{R_e}\right)^{1/2} + 0.34\left(R_e^{0.06} + \frac{1}{1.72 + 0.018\, R_e}\right)\right]^2 \\ 700 \le R_e \langle 10443\end{gathered} \tag{4.17}$$

Brauer[25]

$$\begin{gathered}C_d = \frac{24}{R_e} + \frac{3.73}{R_e^{1/2}} - \frac{4.83 \times 10^{-3} R_e^{1/2}}{1 + 3 \times 10^{-6} R_e^{3/2}} + 0.49 \\ 10443 \le R_e \langle 1.5 \times 10^5\end{gathered} \tag{4.18}$$

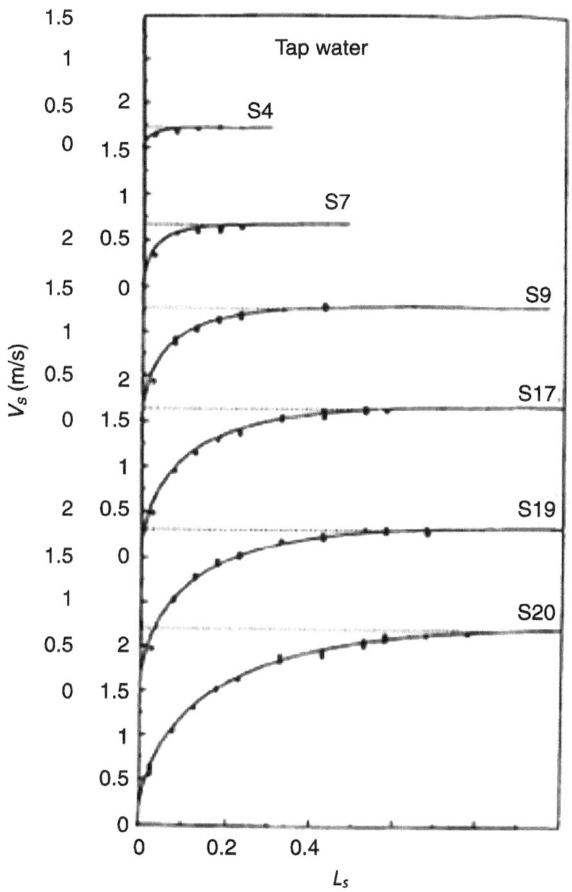

Figure 4.32 ***Fall velocity of a sphere in water.***

For $R_e \geq 1.5 \times 10^5$, the experimental data are compared with Wieselsberger's experimental results.[26]

Fig. 4.32 illustrated the experimental results[7] for the fall velocity in tap water. Solid lines in these figures are the calculated results for Eq. (4.14). Dotted lines denote the 98% fall velocity. Experimental data for tap water agree well with the solid lines.

Experimental data on fine solid particle suspensions are shown in Fig. 4.33. Because non-Newtonian viscosity of these suspensions is neglected for these fluids since it is a dilute mixture, the fall velocity is calculated by substituting the kinematics viscosity of the suspensions in to Eq. (4.14). It is

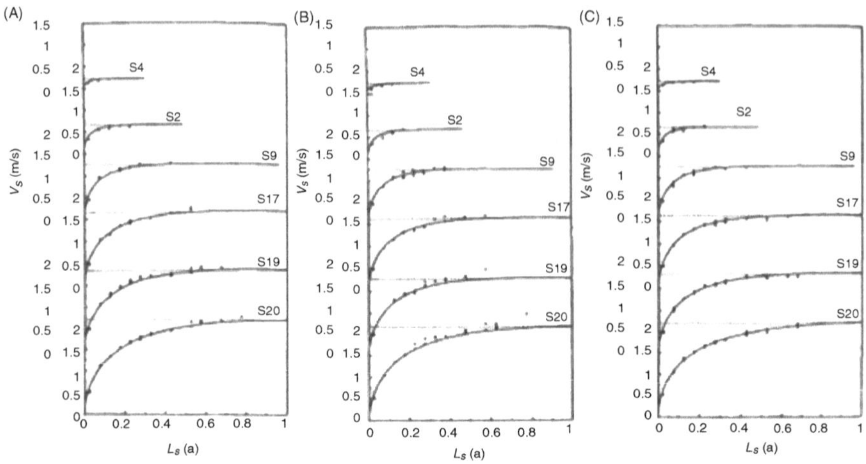

Figure 4.33 ***Fall velocity of a sphere in fine solid particle suspension.*** (A) 0.3% C.B. (B) 0.5% C.B. (C) 0.5% SiO_2.

an interesting result into the carbon black suspension that the fall velocity increases comparing with the calculation result although the data of 0.5% silica dioxide suspension agree well with the solid lines. The phenomenon relates obviously to the drag reduction of flow around a sphere. Also, the onset point of the drag reduction may be represented by the ratio of the fall velocity at the point and the tested sphere diameter, (V_{so}/d_s). The mean value for 0.3% carbon black suspension is larger than that of 0.5% carbon black suspension, and it is 120 and 130 s^{-1} in these suspensions, respectively. It is known that the drag coefficient of a sphere emphasizes the drag reduction, making it clear that the phenomenon occurs in a certain Reynolds number range.

Fig. 4.34A and B represent the relationship between the drag coefficient and Reynolds number. Experimental data of the drag coefficient is calculated by using Eq. (4.14) under the free-fall velocity condition. Dotted and solid lines in these figures indicate the experimental results of tap water and the line given by Lapple and Shepherd[27] for Newtonian fluids, respectively. It is seen in Fig. 4.34A and B that the drag coefficients of 0.1%–0.5% carbon black and 3.0% SiO_2 suspensions at the value of $R_e \geq 8\times10^3$exhibit drag reduction. The value of the drag reduction ratio increases with increasing concentration in the suspension, and maximum obtained value of the drag reduction ratio of carbon black suspension is about 15%. The behavior of the drag coefficient is a like that of dilute polymer solutions that the onset of the drag reduction is $R_e \cong 5\times10^3$.

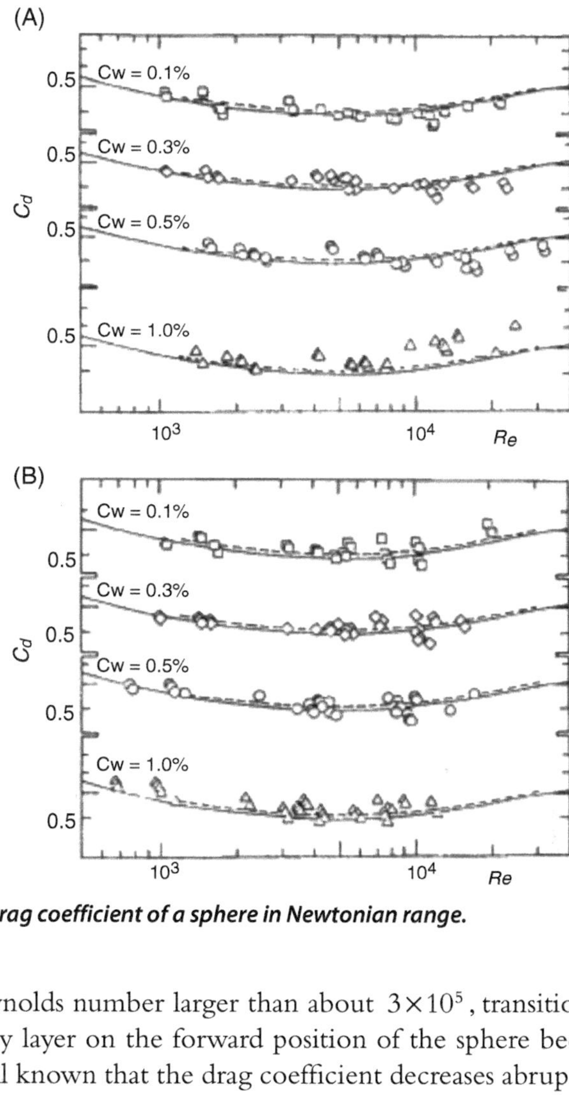

Figure 4.34 ***Drag coefficient of a sphere in Newtonian range.***

At a Reynolds number larger than about 3×10^5, transition occurs and the boundary layer on the forward position of the sphere becomes turbulent. It is well known that the drag coefficient decreases abruptly in a Newtonian fluid, but there are very few studies on the drag of a sphere in water/fine solid particle suspensions above the critical Reynolds number. If the experimental apparatus as shown in Fig. 4.31 is used for the drag coefficient measurement in the critical Reynolds number range, the experimental data are calculated using by the following equation because the data lie in the acceleration range

$$C_d = \left\{(m_s - m_f)g - (m_s + km_f)\frac{dV_s}{dt}\right\} \Big/ \pi\rho_f \frac{d_s^2}{8} V_s^2 \tag{4.19}$$

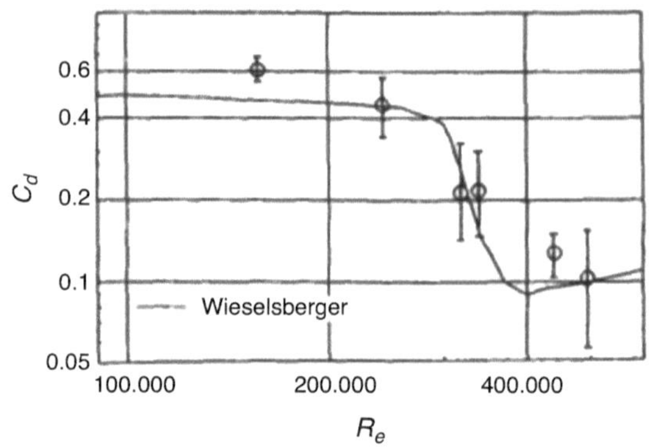

Figure 4.35 ***Drag coefficient of a sphere in tap water for the critical Re number range.***

where the Basset force in Eq. (4.14) is neglected on the assumption that the force is very small at the large-fall-velocity range.

Figs. 4.35 and 4.36A and B are a plot of the experimental data[21] of the drag coefficient of tap water and the suspensions, respectively. The solid lines in graphs represent the experimental results obtained by Wieselsberger. This experiment carried out the measurement for the fall velocity of a steel sphere that it is d_s = 57.1–152 mm in the diameter. The tank shown in

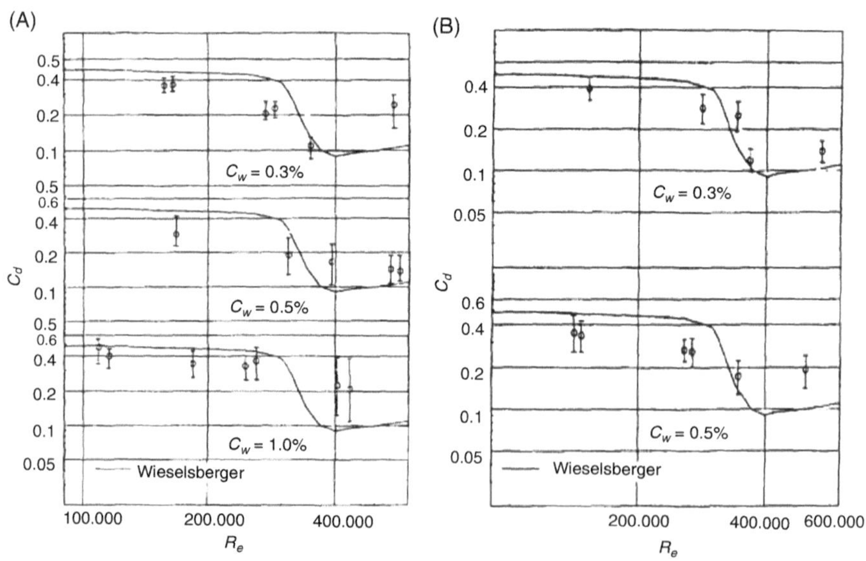

Figure 4.36 ***Drag coefficient of fine solid particle suspensions.*** (A) SiO_2 suspensions. (B) Carbon black suspensions.

Fig. 4.31 are 2200 and 490 mm in the height and the diameter, respectively, was used. Experimental data for water fit almost with Wieselsberger's results.

Fig. 4.36 indicates that for those suspensions, drag reduction occurs in the $R_e \leq 3\times10^5$ range. However, this reduction does not occur above the critical Reynolds number. Since C_d is almost the constant with an increase of R_e, the sudden decrease in the critical range of Newtonian fluids does not occur as shown in Fig. 4.35.

It is known that the surface roughness of a sphere plays a significant role in determining when the reduction in the drag coefficient of Newtonian fluid occurs around the critical Re number range. However, the sudden change does not occur in the case of carbon black suspension. Although the mechanism of the separation was discussed for the roughness and the particle migration from the experimental results of flow in a divergent channel, this fact means that the particle migration is the most influence factor with comparing the roughness effect.

On the other hand, the behavior is similar also to that of dilute polymer solutions.[28–30] And also, as is evident from the experimental results for the separation point of flow around a circular cylinder, this drag reduction is caused by shift in point of the separation. If we assume that the suspension is continuum matter from a macroscopic viewpoint, it can be considered that the drag reduction phenomenon is caused by the visco-elastic effect.

As described above, this pattern of the drag reduction of fine solid particle suspensions is very similar to that encountered in the flow of dilute polymer solutions. Thus, a correlation based on extensive data including the data of other suspensions was presented by refereeing the characteristic time of a boundary layer around a sphere proposed by Hino and Hasegawa[31] for dilute polymer solutions and the relaxation time of fine solid particles.

The characteristic time, T_h and the relaxation time, c are defined as follows:

$$T_h = \left(\frac{d_s}{V_{so}}\right) R_e^{-2} \tag{4.20}$$

$$T_h = \frac{d_e^2}{18\mu}\left(\rho_e - \rho\right) \tag{4.21}$$

where μ and ρ in Eq. (4.21) are the viscosity and the density of water, respectively.

Fig. 4.37 illustrates the relationship between the drag reduction ratio and the nondimensional characteristic time ratio, (T_r/T_h).[7] It is apparent,

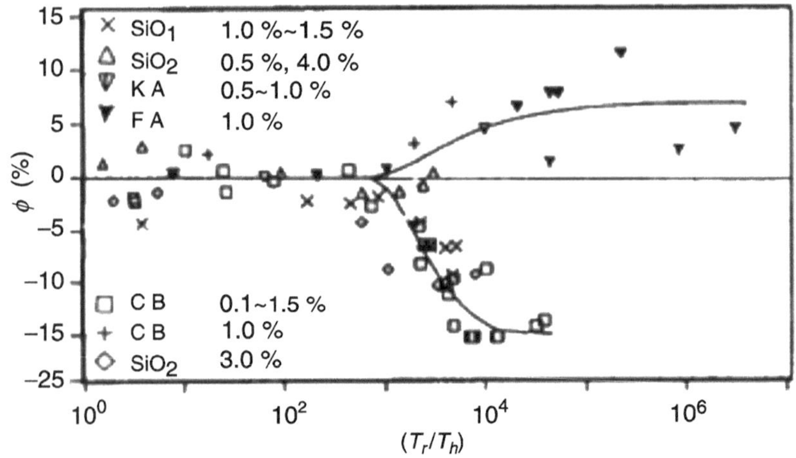

Figure 4.37 ***Effect of (T_r/T_h) on the drag reduction.*** (A) Carbon black suspension. (B) SiO_2 suspension.

in this case at least, that the drag reduction for carbon black or SiO_2 suspension occurs at the value of the parameter above 10^3. This correlation may give onset point of the drag reduction phenomenon although the predictions outlined in Eqs. (4.20) and (4.21) are dependent on the assumed spherical particle.

It is pointed out the equivalent drag reduction phenomenon on a circular cylinder or a sphere for dilute polymer solutions and fine particle suspensions flow. Therefore, it needs to be considered for the viscoelasticity property of the suspensions. Fig. 4.38 shows the measurement instrument for the first normal stress difference. The gap is formed between the cone and the plate, and the clearance angle is $\theta = 1.34$ degrees. It is 40 mm in a diameter of the cone. The thrust is measured by a electric force balance thrust.

The shear rate, $\dot{\gamma}$ and the first normal stress difference, $\Delta\sigma_{n1}$ is calculated as follows,

$$\dot{\gamma} = \omega / \tan\theta \cong \omega / \theta \tag{4.22}$$

$$\Delta\sigma_{n1} = \frac{F}{\pi R^2 / 2} + C\rho_f (R\omega)^2 \tag{4.23}$$

where the second term in the right side of Eq. (4.23) is the correction term for the influence of the centrifugal force. C in the correction term is the

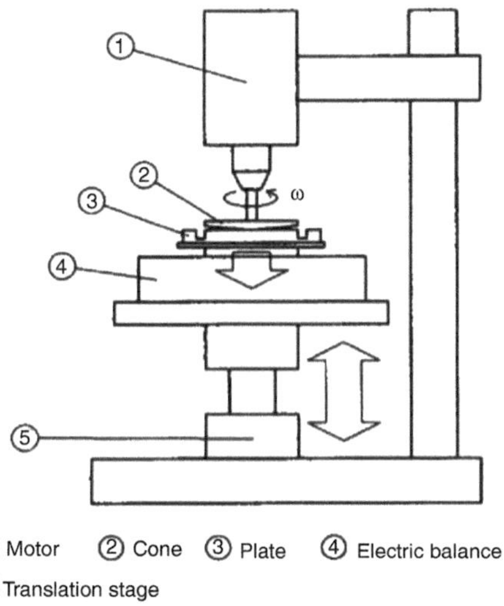

Figure 4.38 ***Cone and plate rheogoniometer.***

function of Reynolds number and is determined from the measurement value of water using Eq. (4.23).

The experimental results for the thrust of carbon black suspensions are indicated in Fig. 4.39. As shown in the figure, the values of thrust of a C_w = 5.0% and a 10.0% suspension increase with the increase of the shear rate and after they decreases. In the case of a C_w = 3.0% suspension, the trend does not appear. The decreasing the measurement value is the influence of the centrifugal force acting on the particles, it is omitted from the

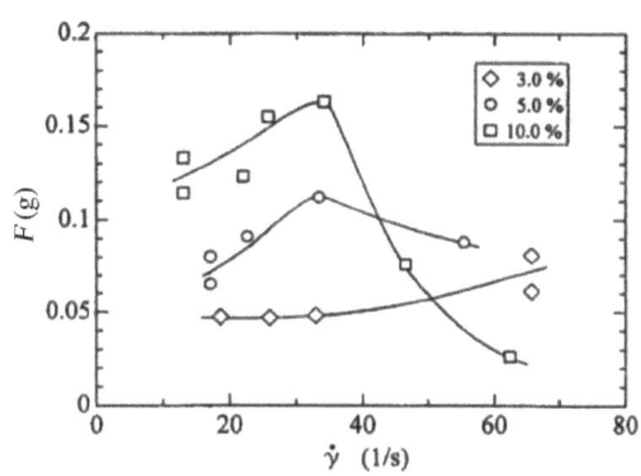

Figure 4.39 ***Relation between the thrust and the shear rate.***

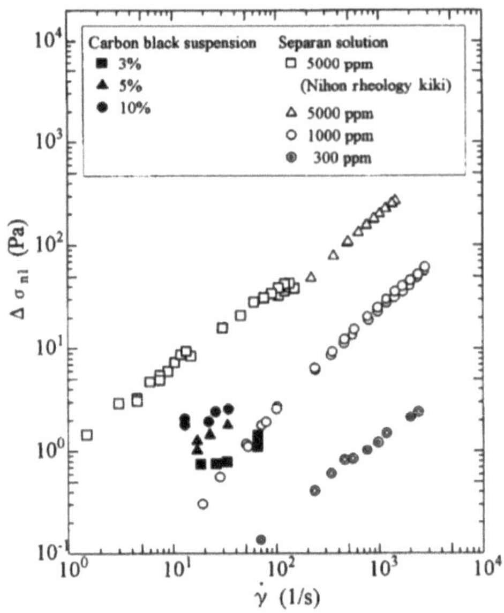

Figure 4.40 ***The first normal stress difference.***

calculation value. The measurement values of the first normal stress difference are shown in Fig. 4.40. The values of Separan solutions are indicated for the comparison in Fig. 4.40. Data show that the value of a $C_w = 3.0\%$ suspension is almost equals with that of a 1000 ppm Separan solution. By using the rheological equation of viscoelastic fluids proposed by Denn,[32] the drag reduction ratio of the drag coefficient of a sphere are arranged. If we choose the diameter, *d* and the terminal velocity, *V* of a sphere as the characteristics length and the characteristics velocity, respectively, Weissenberg number of Denn model is given as

$$W_e = \frac{\lambda}{\mu}\left(\frac{V}{d}\right)^{s-n} \tag{4.24}$$

where μ and *n* are the viscosity and the power related non-Newtonian property. Λ and *s* are the physical constant related the viscoelastic property. In general, $n = 1$ and $s = 2$ are approximated for low concentration range. The first normal stress difference is calculated

$$\Delta\sigma_{n1} = 2\lambda\dot{\gamma}^{s} \tag{4.25}$$

In Fig. 4.40, s of Separan solutions in the range of 300, 1000, and 5000 ppm is given as $s = 0.7$–0.9, the condition of the dilute solutions, $s = 2$ is not satisfied. Because it is very difficult to measure the value of dilute

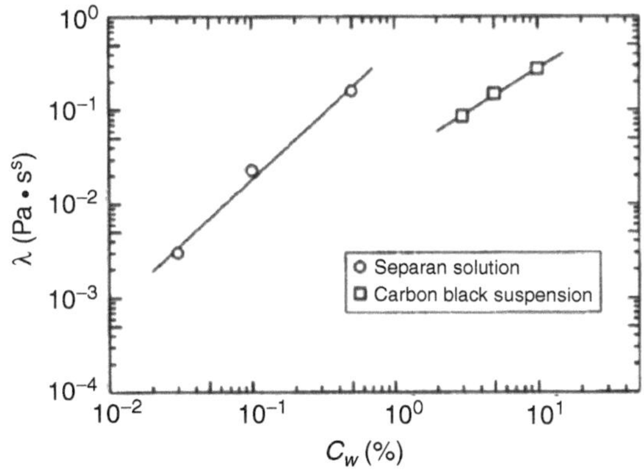

Figure 4.41 ***Relationship between λ and C_w.***

solutions, the values of λ is extrapolated using the experimental data in Fig. 4.40. The results are illustrated in Fig. 4.41. From the graph of Fig. 4.41, it is obtained $\lambda = 1.3 \times 10^{-5}, 8.6 \times 10^{-5}$, and 3.0×10^{-4} Pa s^2 for a 5 ppm, a 20 ppm, and a 50 ppm Separan solution, respectively.

The drag reduction ratio is defined as

$$\frac{\Delta C_d}{C_{dw}} = \frac{C_{dw} - C_d}{C_{dw}} \tag{4.26}$$

where C_d is the drag coefficient of a sphere, and the suffix, w means the value of water.

Using those values, the relationship between the drag reduction ratio obtained in Fig. 4.30 and W_e number are shown in Fig. 4.42. For the

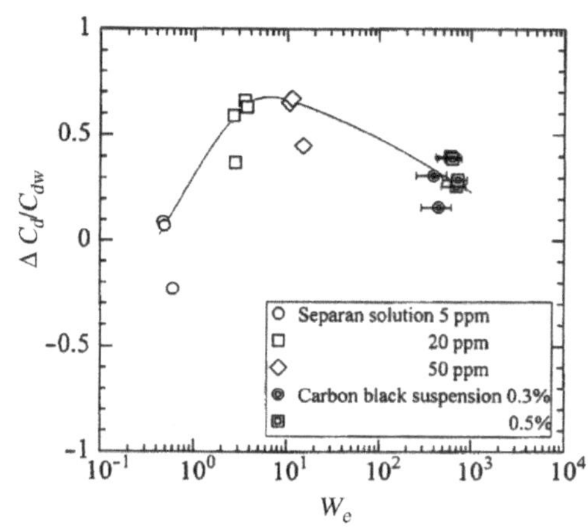

Figure 4.42 ***Relationship between $\Delta C_d/C_{dw}$ and W_e.***

Weissenberg of 0.5, that is, for a 5 ppm separan solution, the drag reduction scarcely occurs. At the Weissenberg number range 4–15, that is, for 20 and 50 ppm solutions the drag reduction occur and the drag reduction ratio is 50%–70% then. Interestingly, the results of carbon black suspensions confirm relevance of the fact that the maximum drag reduction ratio is obtained in the Weissenberg number range.

4.4 MEASUREMENT FOR LIFT ACTING ON A SPHERE

We turn now to a detailed consideration of the particle migration in fluids. The problem with obtaining and interesting data relating to the drag reduction of solid/liquid flow systems is caused by the particle motion in the flow field. We should note, as mentioned above, that a particle in a flow with difference velocity may migrate to the direction that is perpendicular to the flow. In general, the force is defined as a lift. Suffman[33,34] deduced a lift acting a particle, however there is a very few experimental study on the lift. When we analyze the flow behavior of complex mixtures by Lagrange method, it will be necessary to calculation of lift. From this viewpoint, we discuss for lift acting on a particle in nonuniform flow.

To calculate the necessary constants related lift, a flow past a sphere in a rotating vessel is available. We may be able to calculate the lift acting on a particle by using a equation delivered from equilibrium of force.[35] Experimental apparatus[36] to realize it is shown in Fig. 4.43A and B. The rotating vessel in Fig. 4.43A is rotated by a motor and the rotating speed is measured by a revolution indicator. If a sphere having density small than a liquid is put in the rotating vessel filled with a liquid, the sphere moves with rotating liquid and rests under the dynamic equilibrium conditions. The rest point of sphere is measured using a scale plate. The flow pattern of the sphere that is fixed with respect to the rotating vessel is visualized using the flow visualization system with a 300W Xenon lamp as shown in Fig. 4.43B. The tracer is used a hollowness glass beads that it is 20 μm in diameter. Dimensions of test sphere are listed in Table 4.4. Fig. 4.44 shows the experimental results of the equilibrium radius.

The motion equation for a sphere is given as following equation,

$$m_p \frac{d\mathbf{U}_p}{dt} = \mathbf{F}_1 - \mathbf{F}_2 + \mathbf{F}_3 + \mathbf{F}_5 - \mathbf{F}_6 \tag{4.27}$$

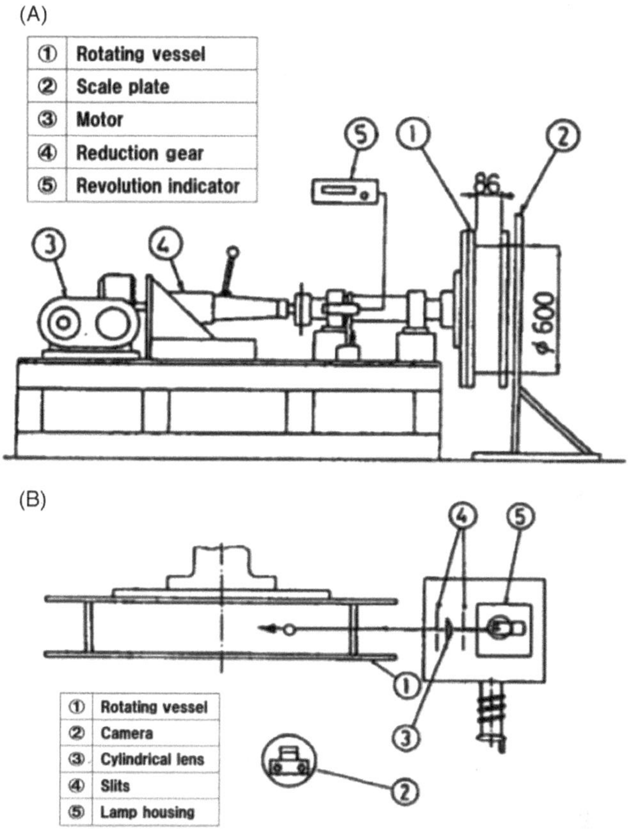

Figure 4.43 ***Experimental apparatus.*** (A) Main unit. (B) Flow visualization system.

Table 4.4 Dimensions of test spheres

Material	Radius α (mm)	Density ρ_s (kg/m³) × 10⁻²
Polypropylene (P.P)	7.93	8.79
	3.96	8.02
Polymethylpentene (P.T.X)	4.75	7.86

where F_1–F_6 are a buoyancy, a drag, a lift a pressure force, Basset force. These forces expect the lift are given as follows,

$$\mathbf{F}_1 = (m_L - m_p)\mathbf{gs} \tag{4.28}$$

$$\mathbf{F}_2 = \frac{\pi a^2}{2} C_d \rho_L (\boldsymbol{U}_p - \boldsymbol{U}_L) \left| \boldsymbol{U}_p - \boldsymbol{U}_{LL} \right| \tag{4.29}$$

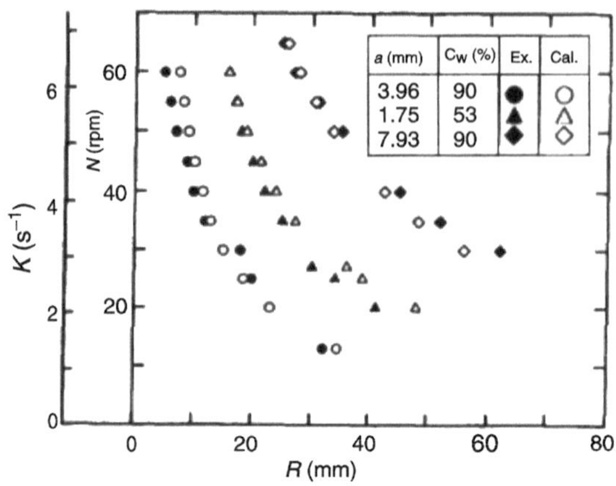

Figure 4.44 ***Equilibrium radios.***

$$\mathbf{F}_4 = \rho_L V_p \left(\frac{D\mathbf{U}_L}{Dt} - \frac{\mu}{\rho_L} \Delta \mathbf{U}_L \right) \tag{4.30}$$

$$\mathbf{F}_5 = C_d \rho_l V_p \left(\frac{d\mathbf{U}_p}{dt} - \frac{D\mathbf{U}_L}{Dt} \right) \tag{4.31}$$

$$\mathbf{F}_6 = 6a^2 \sqrt{\pi \rho_L \mu} \int_0^t \left(\frac{d\mathbf{U}_p}{dt} - \frac{D\mathbf{U}_L}{Dt} \right) \frac{1}{\sqrt{t-\tau}} d\tau \tag{4.32}$$

When the rotation of the vessel becomes the steady state, a particle comes to rest in the first quadrant by the condition of keeping equilibrium between these forces that are the buoyancy, the drag and the centripetal force with a lift.

$$\boldsymbol{F}_1^2 = (\boldsymbol{F}_3 + \boldsymbol{F}_4)^2 + \boldsymbol{F}_2^2 \tag{4.33}$$

Fig. 4.45A–D show the flow visualization results obtained by the apparatus. These spheres remain stationary in the first quadrant. The vortex occurs from the high-speed side of a sphere, and it is no twin vortex in the range of $73 \leq R_e \leq 150$. According to the increase of R_e, an asymmetric twin vortex is formed as shown in Fig. 4.45D.

The experimental results suggest a fact that we can make the computational evaluation for lift by a calculation for Eq. (4.27). Eq. (4.27) is numerically analyzed using Runge-Kutta method. In the analysis, the angular velocity of liquid assumed as $\Omega = Ke^{-1/t}$, and the value fixed for 99% value of K. The drag

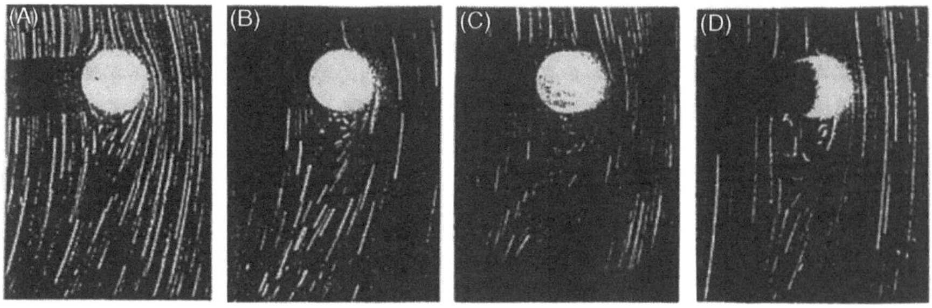

Figure 4.45 ***Flow around a sphere in a rotating vessel.*** (A) $R_e = 40.0$ ($a = 7.93$ mm). (B) $R_e = 73.5$ ($a = 7.93$ mm). (C) $R_e = 153.5$ ($a = 7.93$ mm). (D) $R_e = 184.1$ ($a = 4.75$ mm).

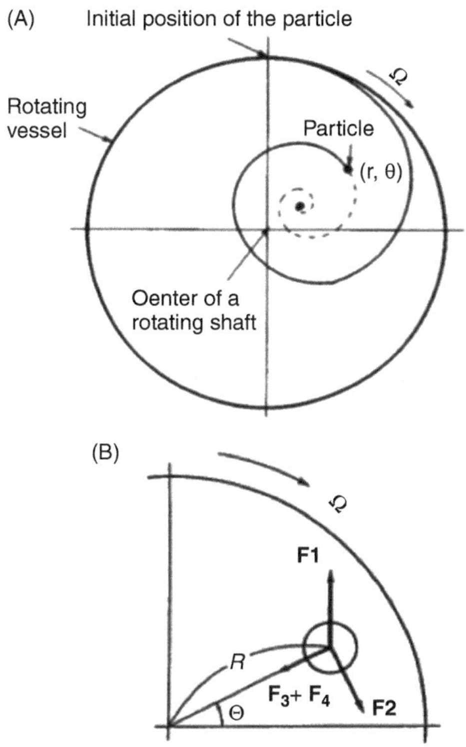

Figure 4.46 ***Particle trajectory and the dynamic equilibrium.*** (A) Sketch of the particle motion. (B) Dynamic equilibrium for the static condition.

coefficient of a sphere is calculated by Eqs. (4.15) and (4.16). Fig. 4.46A and B show the sketch of the trajectory and the dynamic equilibrium at resting state. As point out above, Eq. (4.33) is satisfied in Fig. 4.46B.

An example of the numerical simulation results that a lift force acting on a sphere is neglected, $\boldsymbol{F}_3 = 0$, is illustrated in Fig. 4.47. The results is divided

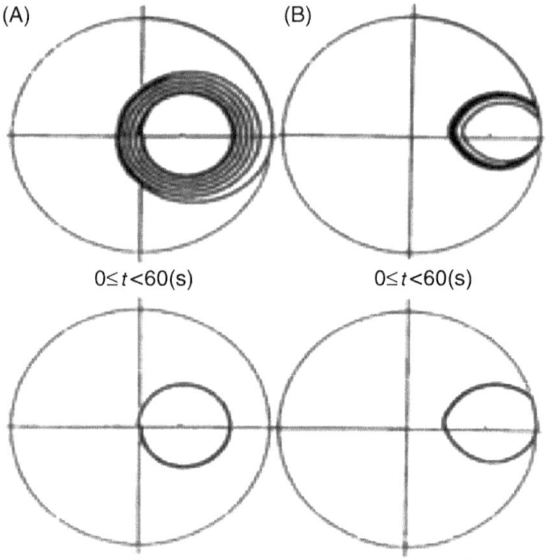

Figure 4.47 ***Calculation results with no-lift condition.*** (A) C_w = 90wt%. (B) C_w = 53wt; a = 7.93 mm.

into halves for the computation time to facilitate visualization. C_w is the concentration of glycerin solutions.

These figures show that the result does not converge within the angular velocity of fluids is a constant value although the experimental result shows the sphere comes to rest under the condition. In other word, we cannot neglect the lift force in this case. Fig. 4.48 shows the result obtained by counting the lift force[33] as follows,

$$F_3 = C_{LS} a^2 r \left| \frac{d\theta}{dt} - \Omega \right| \sqrt{\rho_f \mu \left| \frac{d\theta}{dt} - \Omega \right|} \tag{4.34}$$

where C_{LS} is the lift coefficient given by Saffman.[34] In the calculation, it is assumed that the value of C_{LS} is used 6.46 and the force acts in the direction toward the center. Although the convergence radius decreases in comparison of the results in Fig. 4.47, it does not converge. It can be considered that the application Reynolds number range of Eq. (4.34) is the very small.

Thus, the calculation for a particle trajectory is examined taking of lift defined as the following equation,

$$F_3 = \frac{1}{2} C_L \pi a^2 \rho_f r^2 \left| \frac{d\theta}{dt} - \Omega \right|^2 \tag{4.35}$$

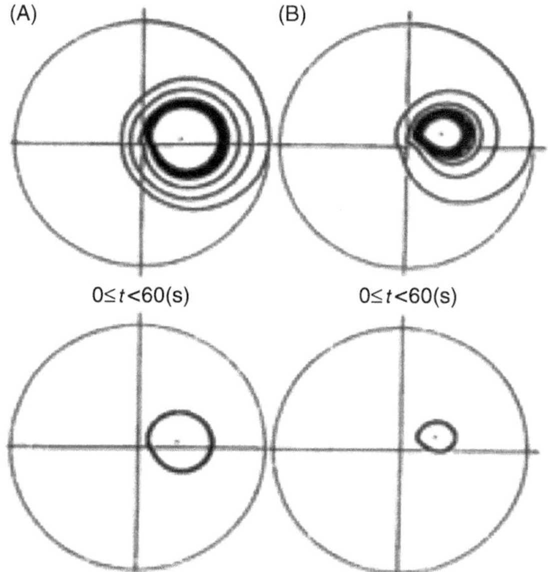

Figure 4.48 ***Calculation results with lift condition.*** (A) $C_{LS} = 6.46$. (B) $C_{LS} = 20.0$.

where C_L is a lift coefficient.

As described above the lift is accounted as the force toward to center. In other words, the direction of action agrees with the direction where the velocity decreases. The calculation was carried out by try and error giving the value of lift coefficient until the solution converges.

An example of the results of having converged is shown in Fig. 4.49. It is seen that the particle reaches to rest in the first quadrant of vessel. These results are given the relationship between the lift coefficient obtained from the simulation and Reynolds number as shown in Fig. 4.50.

As the result, the lift coefficient[36] is deduced as

$$C_L = 7.4 R_e^{-1.2} \quad (6 \times 10^{-2} \le R_e \le 7) \tag{4.36}$$

where R_e is Reynolds number defined as $R_e = 2ar\left|\left(\frac{d\theta}{dt} - \Omega\right)\right| / \nu$. There is a few experimental data on the lift coefficient of a cylinder in nonuniform flow. Yamamoto et al.[37] reported the experimental data for the velocity gradient of linear shear flow at high Reynolds number range using a pendant

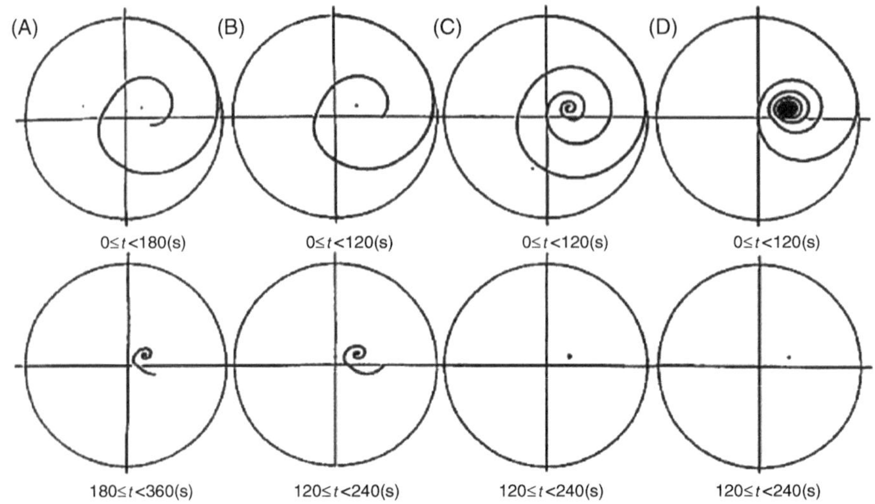

Figure 4.49 ***Calculation results of particle trajectory.*** (A) $C_L = 263$, $R_e = 0.05$, $K = 0.07\ s^{-1}$, $a = 1.1$ mm; (B) $C_L = 105$, $R_e = 0.11$, $K = 0.07\ s^{-1}$, $a = 1.1$ mm; (C) $C_L = 17$, $R_e = 0.5$, $K = 0.2\ s^{-1}$, $a = 2.3$ mm; (D) $C_L = 5.9$, $R_e = 1.2$, $K = 0.02\ s^{-1}$, $a = 3.2$ mm.

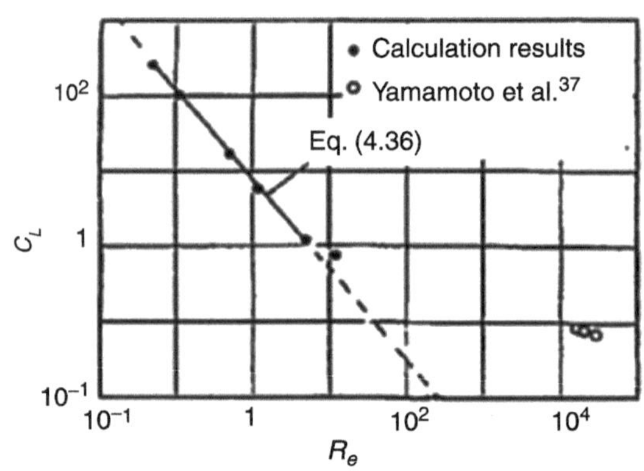

Figure 4.50 ***Lift coefficient obtained from the computational results of particle trajectory.***

method for airflow. In Fig. 4.48, these data are plotted for the comparison of Eq. (4.36). This information about the lift coefficient will becomes an important issue for the numerical simulation of two-phase flow. And also, the experiment for the intermediate Reynolds number region will become a target in future.

Now when a sphere keeps a stationary condition in the flow field, Eq. (4.33) becomes the following equation

$$\left\{g\left(m_L - m_p\right)\right\}^2 = \left(\frac{1}{2}C_d \pi a^2 \rho_L R^2 K^2\right)^2 + \left\{\frac{1}{2}C_L \pi a^2 \rho_L R^2 K^2 + \left(m_L - m_p\right) R\Omega^2\right\}^2 \tag{4.37}$$

where R is a equilibrium radios. The comparison between the experimental and the calculated result for the lift coefficient obtained Eq. (4.25) is shown in Fig. 4.44. The calculation results agree with the experimental results. Experimental results assist in our consideration of particle migration in flow with no uniform velocity.

According to the decrease of the size of a solid particle in complex mixtures, the analyzable range by pseudo-plastic flow is expanded. It is an interesting into the drag reduction phenomenon that a certain type of liquid-solid suspensions has a normal stress similar that of polymer solutions. The experimental data on drag reduction of an external flow show that the separation point moves to the down stream by an increased instability.

Other hand, it will be an important to discuss on the migration of particle in nonuniform flow, and we cannot neglect the influence of lift acting on a particle in the numerical simulation. Pseudo-homogeneous flows fall into an intermediate area for the analysis.

REFERENCES

1. Daily JW, Bugilarello G. *Tappi* 1961;**44**(7):497.
2. Mih W, Parker J. *Tappi* 1967;**50**(5):237.
3. Kato H, Mizunuma H. *Bull JSME* 1983;**26**(212):231 (1st Report, Pipe flow).
4. Watanabe K, Ogata S. *Proc World Academy Sci Eng Tech* 2013;**82**:1.
5. Fiber Handbook. The Society of Fiber Science and Technology, Japan. Maruzen, 1994, Pl.2, p.15.
6. Virk PS, Chen RH. In: *Proceedings of the 2nd International Symposium on Seawater Drag Reduction*; 2005p. 545.
7. Watanabe K, Ohira H, Kato H. In: *Proceedings of 1st ASME–JSME Fluid Engineering Conference*; 1991p. 119 FED-Vol. 110.
8. D.W. Dodge and A. B. Metzner, Paper presented at Annual American Institute of Chemical Engineers Meeting (1957).
9. S. Ogata, et al., Proceedings of JSME'16 Fluid Engineering Division Meeting (in Japanese), (2016), p. 1.
10. Sellin RHJ, et al. *J Hydraul Res* 1982;**20**(3):235.
11. Verkerk CG. *Hydraulictransport* 1982;**8**:307 F4.
12. Watanabe K, Iwata T, Kato H. *Bull JSME* 1984;**27**(230):1682.

13. Watanabe K, Kamoshida T, Kato H. *Experimental heat transfer, fluid mechanics and thermodynamics*. Elsevier Science; 1988; 693.
14. Mizushina T, Mitsuishi N, Nakamura K. *Trans Soc Chem Eng* 1964;**28**(8):648 Japan (in Japanese).
15. Watanabe K, Feng C, James DF. *Int J Multiphase Flow* 2000;**26**:729.
16. A. Fage and J. H. Warsap, ARC R & M (1930), p. 1283.
17. Schlichting H. *Boundary layer theory*. New York: McGraw-Hill; 1968.
18. Dryden HA, Hutton JF, Walter K. *J Aero Sci* 1953;**20**:477.
19. Leighton D, Acrivos A. *J Fluid Mech* 1987;**181**:415.
20. Watanabe K, Chang Y, Fujita T. *Trans JSME (in Japanese)* 1996;**62**(603):3818.
21. Watanabe K, Kui H. *ASME, IEAFM, FED* 1995;**221**:127.
22. Asakura K, et al. *J Mining Met Inst Japan (in Japanese)* 1983;**99**(1064):457.
23. Schiller L, Nauman A. *Z Ver Deut Ing* 1933;**77**(12):318.
24. W. L. Carey, Ph.D. thesis, Syracuse University, New York (1970).
25. Brauer H. *Chemic-Ing Tech* 1973;**45**:1099.
26. Schlichting H. *Boundary-layer theory*. 7th ed. McGraw Hill; 1979 p.15.
27. Lapple CE, Shepherd CB. *Ind Eng Chem* 1940;**32**(5):605.
28. Ruszczycky MA. *Nature* 1965;**206**:614.
29. White DA. *Nature* 1966;**212**:177.
30. White A. *Nature* 1967;**216**:994.
31. Hino M, Hasegawa S. *Tech Rep Depart Civil Eng Tokyo Inst Technol (in Japanese)* 1968;**6**:123.
32. Denn MM. *Chem Eng Sci* 1967;**22**:395.
33. Suffman PG. *J Fluid Mech* 1965;**22**(Part 2):385.
34. Suffman PG. *J Fluid Mech* 1968;**31**:624.
35. Fisher AL. *Ins Cont Eng* 1962;125.
36. Watanabe K, Sugimura Y, Yaita K. *Trans JSME (in Japanese)* 1996;**62**(603):3804.
37. Yamamoto F, et al. Trans JSME (in Japanese) 1991;57(542):3414.

CHAPTER 5
Drag Reduction in a Homogeneous Flow

Contents

5.1 Introduction 123
5.2 Polymer Solutions 124
  5.2.1 Flow in a Pipe 124
  5.2.2 Flow Between Two Coaxial Cylinders 130
  5.2.3 Flow Around a Rotating Disk 139
  5.2.4 Flow Past a Sphere 152
5.3 Surfactant Solutions 157
  5.3.1 Flow in a Pipe 157
  5.3.2 Flow Between Two Coaxial Cylinders 166
  5.3.3 Flow Around a Rotating Disk 169
  5.3.4 Flow Past a Circular Cylinder 179
5.4 Biopolymer Solutions 194
References 203

5.1 INTRODUCTION

Many studies have been experimentally or analytically conducted about drag reduction in the flow of dilute polymer solutions for a long time. High polymer constitution is the molecules, which a monomer was tied up to by covalent bond, and it becomes the stick or coiled molecules by the combination with the water molecule with aqueous solution with increase in molecular weight and changes in structure. Such a polymer structure is changed by mechanical shearing, which is agitated or transported by a pump. As a result, it causes the decrease of the amount of drag reduction performance. If the condition is retained, the drag reduction disappears. The degradation of polymer solutions has been considered to be major cause to obstruct industrial application. On the other hand, the drag reduction phenomenon occurs even in flow of surfactant solutions and it has been applied to the heating medium transport system as the degradation disappears under such condition. Under shearing condition, a rod like micelle is formed in liquid and the viscosity increases with the decreasing the shear rate. The difficulty remains in terms of the size and the shape decisions about polymer or surfactant in liquid.

Drag Reduction of Complex Mixtures
http://dx.doi.org/10.1016/B978-0-12-809920-9.00005-6

In this chapter, it has an interest in flow of polymer or surfactant solutions from a macroscopic viewpoint, and some unique phenomena in turbulent flow are demonstrated using the experimental results for not only a circular pipe flow but also a flow between two coaxial circular cylinders and a flow around a sphere or a circular cylinder. The reader may be able to find a problem that remains to be solved in these experimental results in future.

5.2 POLYMER SOLUTIONS

5.2.1 Flow in a Pipe

Drag reduction phenomenon about high molecular polymer flow is well known as Toms effect[1] that dilute polymer solutions occur drag reduction phenomenon for turbulent flow in a circular pipe. Many experimental and analytical results for it have been reported to date. In laminar flow range, the phenomenon does not occur as there is no turbulence in the range. If we want to get drag reduction in laminar flow range,[2] it should be noticed that fluid slips on the solid surface. In such a case, it may be able to expect that we can obtain the drag reduction phenomenon in both laminar and turbulent flow ranges.

This result[3] is shown in Fig. 5.1. As non-Newtonian viscosity cannot be ignored for the concentration range of more than 200 ppm, the friction factor is arranged using power law Reynolds number,

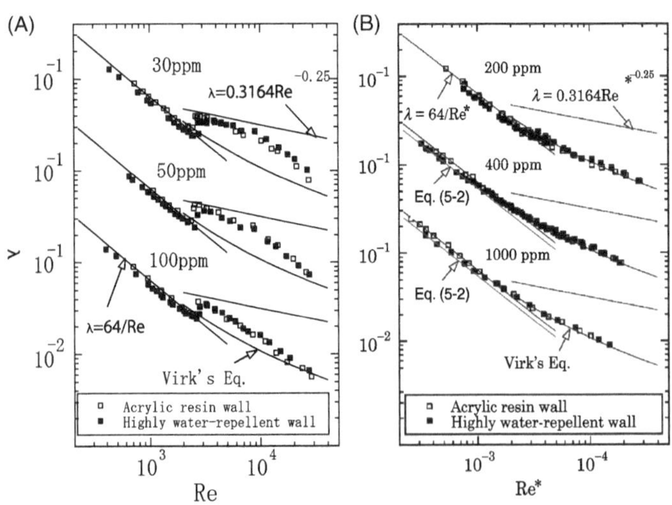

Figure 5.1 ***Friction factor of PEO solutions in a pipe with highly water-repellent wall.*** (A) Dilute solutions and (B) highly concentrated solutions.

$\text{Re}^* = (8n^n \rho d^n U^{-2-n}/2^n(3n+1)^n K)$. Experimental data of an acrylic pipe with no-slip agree well with the conventionally reported results in laminar and turbulent flow[4] range. But, the experimental data of the highly water repellent wall pipe demonstrate the drag reduction in laminar flow range. In other words, we can obtain drag reduction as it occurs in both flow range of laminar and turbulent by using a kind of hydrophobic wall pipe.

If the fluid slip for a power-law fluid, $\tau_{rz} = K\left(-dv_z/dr\right)^n$ is assumed as the following equation by modifying Navier's hypothesis,[5] we can obtain the friction factor analytically.

$$\tau_w = \beta u_s^n \tag{5.1}$$

Friction factor is given by the following equation[3]

$$\lambda = \frac{64}{\text{Re}^*}\frac{1}{\left[1+\dfrac{(3n+1)K^{\frac{1}{n}}}{nR\beta^{\frac{1}{n}}}\right]^n} \tag{5.2}$$

where, it is assumed that n in Eq. (5.1) equals to the exponent of a power-law fluid and Re★ is the power-law Reynolds number. In Fig. 5.1B, the *solid lines* are the calculated results substituting the extrapolation for β in Eq. (5.1). Experimental data fit with the line.

Fig. 5.2 is the rearrangement result of experimental data in Fig. 5.1A. Furthermore, the image of drag reduction phenomenon read from the

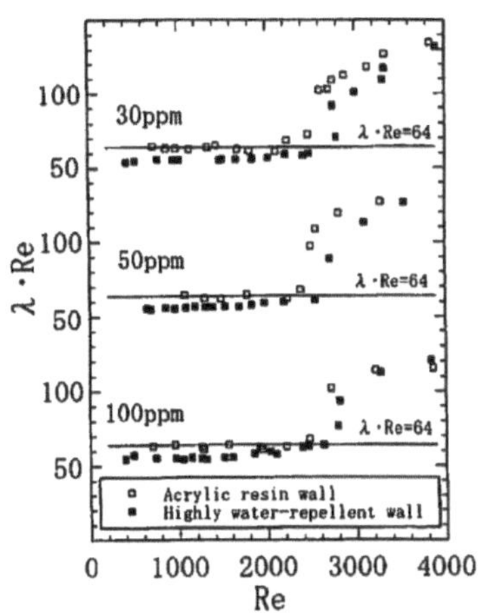

Figure 5.2 ***Rearrangement result of experimental data of Fig. 5.1A.***

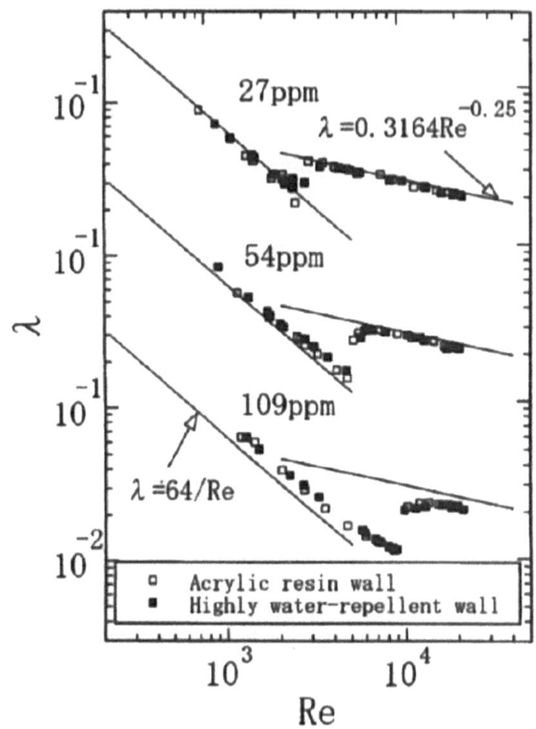

Figure 5.3 ***Friction factor of surfactant solution, Ethoquad O/12.***

graph of Fig. 5.1A is clearly distinguished. That is, it has shown the reduction in laminar flow range, the transition delay from laminar to turbulent with the increase of the concentration and the reduction of the ratio in turbulent flow range. Consequently, we can obtain the drag reduction in both flow ranges in the case of flow in a highly water-repellent pipe. However, the difference does not occur for drag reducing surfactant solutions as shown in Fig. 5.3. Experimental data of an acrylic resin pipe agree well with that of a highly water-repellent pipe. It stems largely from declining surface tension of the solutions and the surface structure.

The behavior of dilute polymer solutions in a micro tube should be noticed as a unique phenomenon for laminar drag reduction. It is a phenomenon that the flow rate of dilute polymer solutions in a micro tube increases compared with the flow rate calculated by substituting the viscosity value, which was measured by means of a rotational cone viscometer. The results[6] are shown in Fig. 5.4A–C.

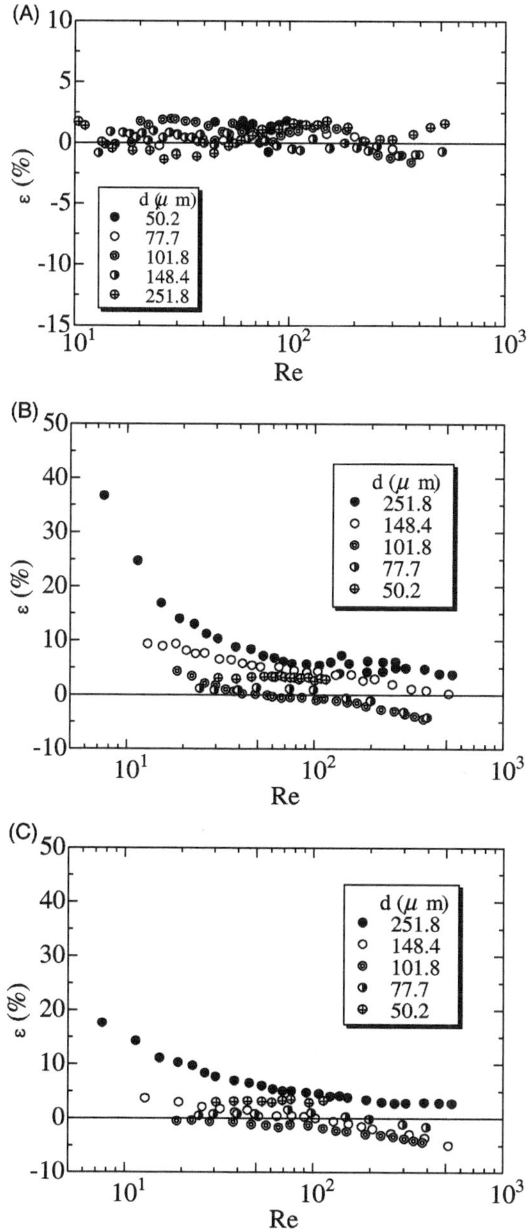

Figure 5.4 ***Flow rate increasing ratio.*** (A) Distilled water, (B) Peo 5 ppm solutions, and (C) Separan 3 ppm solutions.

$$\varepsilon_{error} = \left\{ \left(Q_{ex} - Q_{ca} \right) / Q_{ca} \right\} \times 100(\%) \tag{5.3}$$

where Q_{ex} and Q_{ca} are the measured and calculated flow rate in a micro tube, respectively. Q_{ca} is calculated by applying Hagen–Poiseuille equation, because the non-Newtonian viscosity can be neglected for less than 5 ppm and 3 ppm of the concentration in Peo and Separan solutions, respectively.

In Newtonian case, the data of the increasing ratio lay in the range of ±0.2% in the maximum as shown in Fig. 5.4A. But, the ratios of dilute polymer solutions in Fig. 5.4A and B increase in the range of the low Reynolds number range. The increasing of the flow rate occurs in the range of $Re < 50$ for $d = 50.2 \sim 251.8\ \mu m$ in the case of Peo 5 ppm solution, and the increasing ratio decreases with the decreasing of the tube diameter. It has a minimum value at the case of $d = 101.8\ \mu m$.

It is pointed out that the increase in the flow rate occurs by the apparent fluid slip and is assumed, in general, that it depends on a formation of a thin layer of polymer solutions near tube wall. In other words, the fluid slip occurs apparently near the wall because the viscosity of the layer is a small value comparing with that of the flow in a tube. The apparent slip velocity of polymer solutions increases with the increase of the wall shear rate. However, it decreases for the rate of the total flow rate because the decrease in the viscosity of the shear thinning presents remarkably. For the effect of the tube diameter on the viscosity, Quibrahim[7] clarified the apparent viscosity decreases with the decreasing of the diameter and it increases over the critical value of the diameter.

On the other hand, Muller-Mohnssen[8] clarified the deviation of the flow by measuring the velocity profile of a micro-channel. But, the unbalanced thin layer, which may cause the apparent fluid slip, is not confirmed yet experimentally. The problem complicates because the polymer network in view point of microstructure, the adhesion to the wall surface and the shear thinning phenomena effect on it. To simplify the problem, it will be necessary to study for very dilute solutions does not show the shear thinning effect in the flow.

Although there is unknown factor to clarify the mechanism for the increasing flow rate in laminar flow of dilute polymer solutions, it may be necessary to measure the velocity profile in a micro channel. Watanabe et al. carried out the measurement of the velocity profile in a micro channel by using a Micro PIV system. The system is shown in Fig. 5.5. The width and the height of the section of the micro channel are $w = 160\ \mu m$ and $h = 43\ \mu m$, respectively. The wall of the micro channel is made of plexiglas

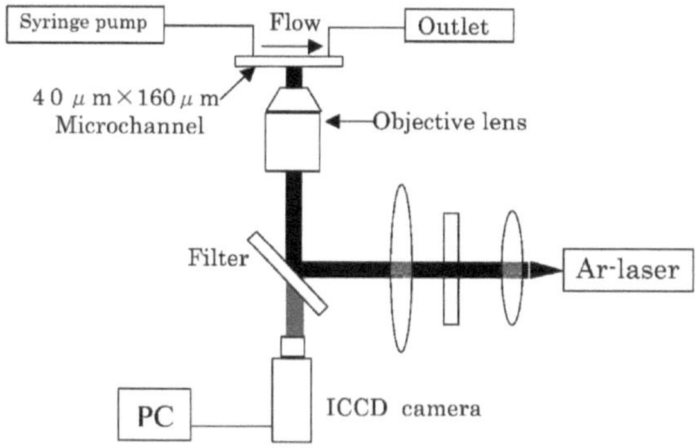

Figure 5.5 ***Micro PIV system.***

and silicon. Test fluids are distilled water and Peo 5 ppm solutions. The tracer is the fluorescent particle that it is 1 μm in the diameter and the tracer was visualized by the laser excitation. Test fluids are run into the micro channel with a constant flow rate by a syringe pump. The visualization image was filed in a personal computer after 130 sheets were photographed in the 1/30 s interval by a CCD camera. The velocity vector of the flow was obtained by the low-density PIV analysis that the preservative image data processing is done by using the Dipp-Flow software. The measuring Reynolds number is $Re = 6.6 \times 10^{-3}$. The value is very small compared to when measuring the pressure loss data, however it is no essential difference because the flow is laminar and the flow system size of this experiment is 160 μm.

The experimental data of the velocity profile in the micro channel are shown in Fig. 5.6. The *solid line* in the figure shows the analytical velocity profile[9] given by the following equation for a Newtonian fluid flow:

$$v = \frac{y}{2\mu}\left(-\frac{dp}{dz}\right)(h-y) - \frac{4h^2}{\pi^3\mu}\left(-\frac{dp}{dz}\right) \sum_{n=1,3,5,\cdots}^{\infty} \frac{1}{n^3}\sin\left(\frac{n\pi}{h}y\right)\frac{\cosh(n\pi x/h)}{\cosh(n\pi w/2h)} \tag{5.4}$$

where h and w are the height and the width of the channel, respectively. (See Example 2.1 in Chapter 2 for the derivation of Eq. (5.4)) The data are the experimental results at the center of the channel z = 21.5 μm. $x = 0$ and $x = 160$ μm denote the positions of the side wall of the channel. Experimental data of distilled water agree well with the solid line. In the case of Peo 5 ppm solutions, the unique velocity profile is not recognized in the result although Muller-Mohnssen et al. reported the remarkable change

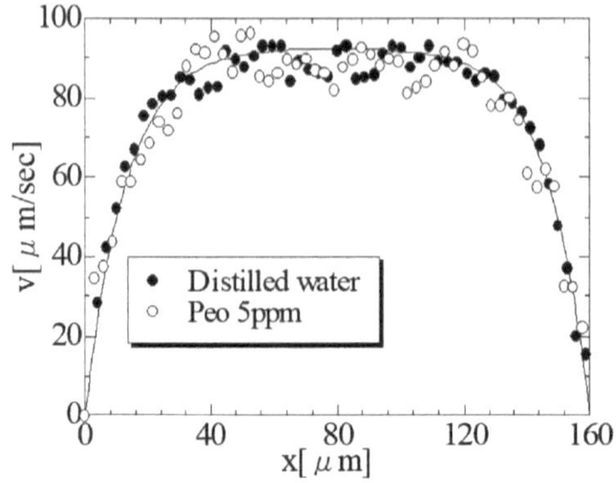

Figure 5.6 ***Velocity profile of polymer solutions in a micro-channel.***

of the velocity profile in 0.25% anion polyacrylamide solutions. It is 3 μm in the resolution of the measurable distance for the Micro PIV system in this study. It may be considered that a unique velocity profile of dilute polymer solutions occurs in the range of fewer than 3 μm because this fact is analogized also from the previous study[10] that it is about 0.8~1.5 μm in the adsorption thickness of high molecular polymer. Thus, a supposition that the increase of the flow rate is occurred by the existence of polymer thin layer at the wall because the no-slip boundary condition was satisfied at the wall, may be generally concluded.

5.2.2 Flow Between Two Coaxial Cylinders

In general, the flow pattern between two coaxial cylinders changes considerably with speed, because of the instability. It is known that the pattern can change from a rotational Couette flow to a Taylor–Couette flow with an increase in speed, for a Newtonian fluid between a rotating inner cylinder and a stationary outer cylinder. There are two combinations of the rotation for the flow system; this is a case of flow of a rotating inner cylinder and stationary outer cylinder, and another case is flow of a stationary inner cylinder and a rotating outer cylinder. The former becomes unstable flow with an increase of the rotation comparing to the latter. Fig. 5.7 shows the change of flow pattern in the case of a rotating inner cylinder and a stationary outer cylinder with increase in the angular velocity, ω.

This flow field is ideal for research into the transition process from laminar to turbulent flow because it is slower than other flow systems and the instabilities are well understood. Thus, much theoretical and experimental research on the transition process using this flow field has been carried out for Newtonian fluids since Taylor's work.[11] Cross section of Taylor vortex is shown in Fig. 5.8.

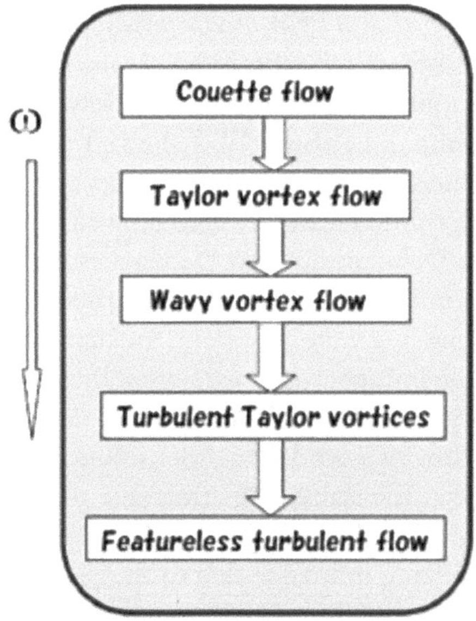

Figure 5.7 ***Flow pattern between two coaxial cylinders.***

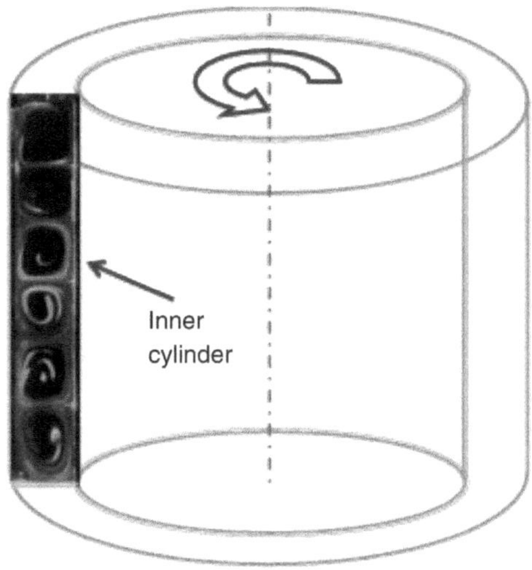

Figure 5.8 ***Taylor vortex in gap of rotating cylinders.***

On the other hand, polymer solution behaves like a viscoelastic fluid. The rheological behavior and flow characteristics depend on the types of polymer additives and their concentrations. From prior studies conducted using polymer solutions, the problem of a purely elastic Taylor-Couette instability has been noted. Larson et al.[12] showed that, for the Oldroyd-B fluid, there exists an inertia-free mode of instability in a Taylor-Couette flow. By means of a linear stability analysis, they showed that secondary toroidal cells occur when the Deborah number De reaches 20, which was in agreement with the experiments with a solution of 1000 ppm high-molecular-weight polyisobutylene in a viscous solvent. Sadanandan and Sureshkumar[13] studied analytically the influence of elasticity on the budgets of the vorticity and the kinetic energy associated with the most dangerous disturbance to the plane Poiseuille flow of a model polymer solution, using the Oldroyd-B fluid. They showed that the stabilizing effect due to the perturbation shear stress increases monotonically with increase in the elasticity number, and also showed a destabilizing influence due to the perturbation normal stress that increases monotonically. Sureshkumar et al.[14] calculated the bifurcating families which correspond to each of the two possible nonaxisymmetric patterns emerging at the point of criticality, namely, the spirals and ribbons, to determine their stability. They showed that the unstable oscillatory modes are always nonaxisymmetric, and transitions to them are probably discontinuous. Although the extensional nature of the fluid does play an important role, the resulting flow patterns in this case were not shown.

Regarding the character of instability in a Couette-Taylor flow, Groisman and Steinberg[15] showed experimentally that two novel oscillatory flow patterns are observed after small additions of polymers. One of the patterns was essentially due to the fluid elasticity, and the other resulted from an inertial instability modified by the elasticity. Lee et al.[16] carried out a flow visualization study to reveal the effect of dilute polymer solutions on the Görtler instability between two cylinders. They showed experimental results in which the Görtler instability was stabilized by the polymer additive, which created increase in the local viscosity.

In general, these works found that viscoelasticity increases the critical Taylor number at which a cellular flow is observed. Thus, an increase in viscoelasticity stabilizes the flow against the formation of initial Taylor vortices. However, the formation process of Taylor vortices has not been clarified experimentally in detail. Although it has been speculated that the transition in a Newtonian fluid is different from that in a viscoelastic fluid, the vortex formation process has not been sufficiently studied.

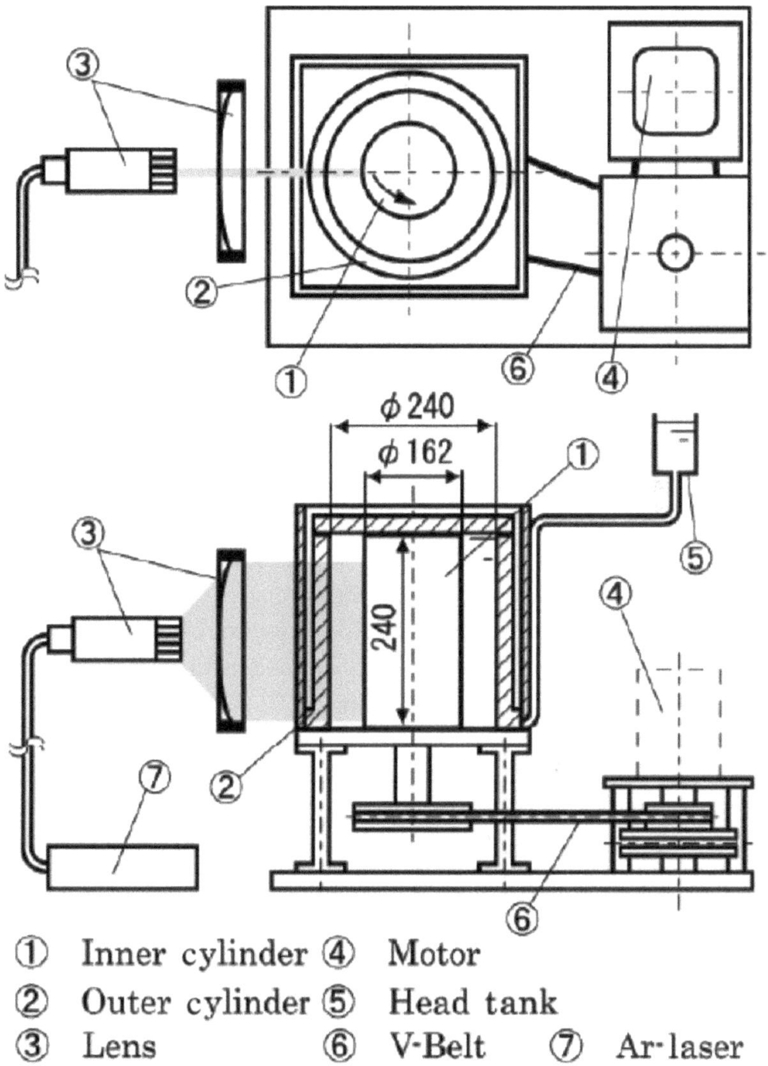

Figure 5.9 ***Experimental apparatus for flow visualization system.***

Watanabe et al.[17] obtained the flow visualization results using the experimental appatatus as shown in Fig. 5.9. It is a two-coaxial-cylinder system in which the inner cylinder is rotated at a constant speed by a servomotor, and the outer cylinder and end wall are stationary. The height and the gap between the cylinders are L = 240 mm and δ = 39 mm, respectively. The radius of the outer cylinder is R_0 = 120 mm, as the aspect ratio $\Gamma = (L/\delta)$ is 6.15, six Taylor vortices should appear in the gap. The angular velocity of the inner cylinder ω was measured by means of a digital counter. The inner and outer cylinders are made of aluminun and acrylic resin, respectively. An acrylic resin container, filled with tap water to protect the light reflected from the laser sheet, covers the outer cylinder. The formation and evolution of counter-rotating vortices close to the inner cylinder wall were examined by means of a laser-induced-fluorescence (LIF) flow visualization technique. A laser beam was directed radially through the center of the cylinder and spread into a vertical leaser sheet by a cylindrical lens, illuminating the flow in the vertical plane. The thickness of the sheet was about 5 mm. The test fluids were marked with a fluorescein dye injected upstream of the laser sheet using a sylinge, via a 1 mm tube inserted between the cylinders through a hole in the top of end plate. The most effecttive method of marking the flow close to the inner cylinder was to slowly bleed dye directly onto the surface of the inner cylinder, and the near-wall vortices entrained the dyed fluid away from to make the vortices visible. It was confirmed, using a capillary viscometer, that there was no change in the viscosity of liquid which was mixed with the visualization dye. Visual records of the LIF experiments were made using a video system. The range of nondimensional Taylor number $T_a^* = (T_a/T_{acrit})$ defined as the ratio of the Taylor number T_a and the critical Taylor number T_{acrit} is 4.6×10^3. The range was $30 \leq T_a^* \leq 236$. Here the Taylor number is defined as $T_a = \left(R_i \omega_i^2 \delta^3\right) / \upsilon^2$ where δ is $(R_0 - R_i)$ and ν is kinematic viscosity.

At first, it will be necessary to clear the forming process about Taylor vortex of tap water to compare with that of dilute polymer solutions or surfactant solutions. Fig. 5.10A–F shows an example of the visualization results on the forming process of Taylor vortex of tap water. As the rotation of inner cylinder, Göltorer vortecies generate from both side of inner cylinder wall, and it forms into the stable cell with varying time. Fig. 5.10F shows that six Taylor cells appear in this flow system as described earlier.

Fig. 5.11 shows the visualization result[18] of Separan 100 ppm solution. After the rotation starts, we can see that the dye streak has separated from the inner cylinder wall as shown in Fig. 5.11A. With an increase in the rotational speed, a Couette flow, as shown in Figure 5.11B, after t = 420 s. The number of streaks almost agrees with the number of rotations. The secondary flow occurs at the corners of the upper and end plates after t = 600 s, and the secondary flow range becomes larger. When the size of the vortex reaches approximately one quarter of the cylinder height, the second vortex, which rotates in the reverse direction to the first vortex, is generated. After t = 1080 s in Fig. 5.11F, we recognize the occurrence of a Görtler

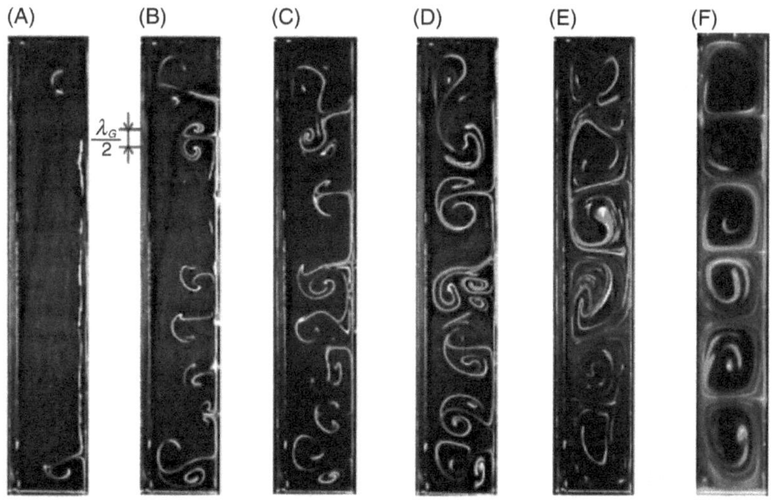

Figure 5.10 ***Formation process of Taylor volrecies in tap water.*** (A) 40 s, (B) 60 s, (C) 70 s, (D) 80 s, (E) 150 s, and (F) 540 s.

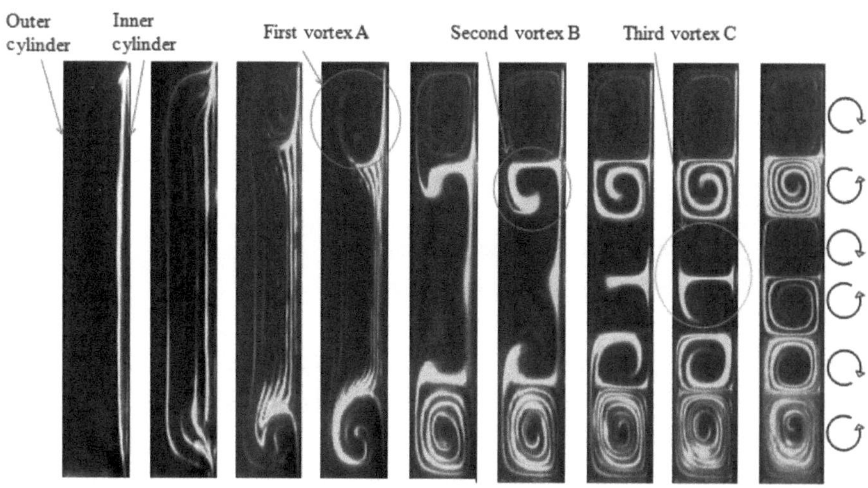

Figure 5.11 ***Forming process of Taylor cell of Separan 100 ppm solutions.*** (A) 180 s, (B) 420 s, (C) 600 s (D) 720 s (E) 1020 s (F) 1080 s (G) 1140 s (H) 1170 s (I) 1260 s.

vortex from the swelling of the dye streak at the center of the cylinder wall. Finally, the six Taylor cells are fixed in the gap, as shown in Fig. 5.11I, after $t = 1260$ s. The flow pattern obtained in this experiment is that of a normal mode of Taylor vortex flow.

In the early stage of the vortex formation, the effect of the concentration on the first vortex is indicated in Fig. 5.12. The axial wavelength of

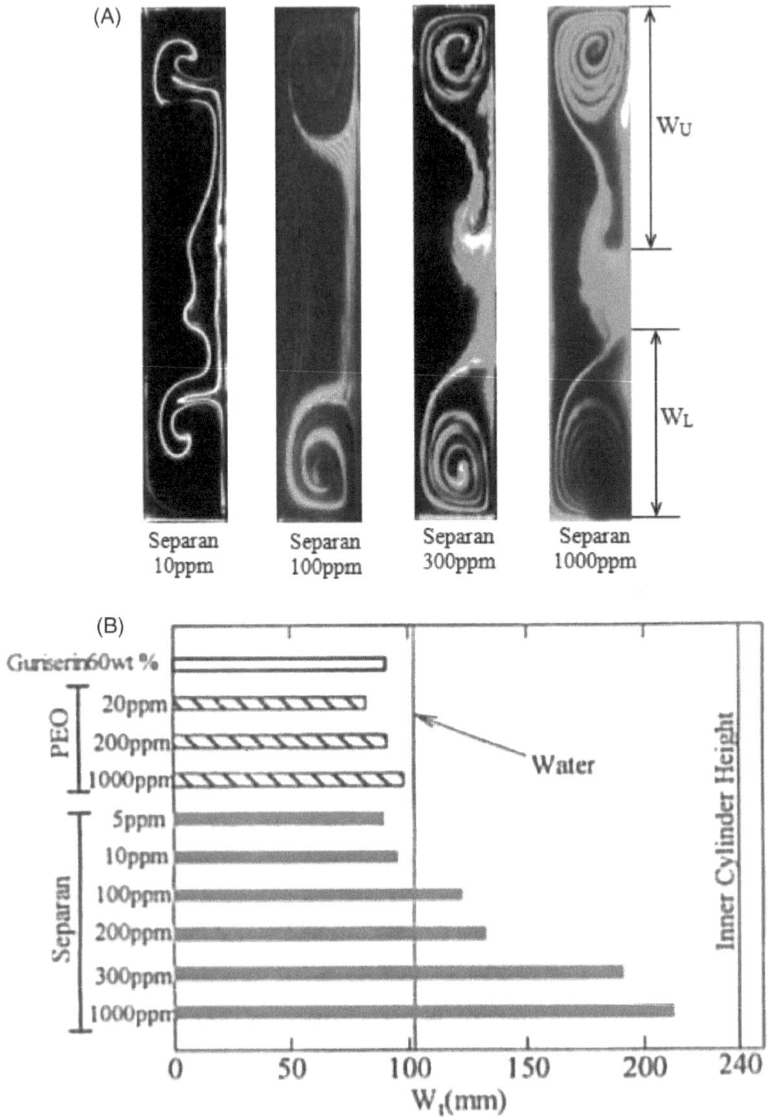

Figure 5.12 ***The early stage of the vortex formation of polymer solutions.*** (A) Effect of concentration and (B) total axial wavelength of the first vortex.

the first vortexes in the early stage becomes larger with increasing polymer concentration. The total wavelength of the Peo and Separan solutions are arranged in Fig. 5.12B. W_i is the total axial wavelength, and the value of W_i

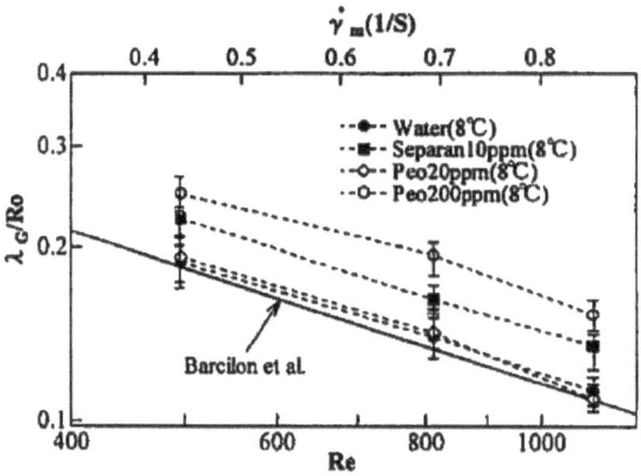

Figure 5.13 ***Gőrtler wavelength of polymer solutions.***

is defined as the wavelength in which the second vortex arises. In the case of Separan solutions, it increases substantially with the increase in polymer concentration. To grasp the effect of polymer additives on the stable Taylor cell node, the axial wavelength of each cell in the stable mode is measured. When the flow pattern depends on the time elapsed from the start of rotation of the inner cylinder, it is important to allow the system to achieve stable mode. In this measurement, using Snyder's formula[19] the stable time was determined to be 3 h from the rotation start time.

Fig. 5.13 shows the effect of polymer concentration on the wavelength of Gőrtler vortex. The acceleration of the inner cylinder is $d\mathrm{Re}^{\star}/dt = 750/\mathrm{min}$. As this acceleration progresses, a fixed Reynolds number is achieved within 1 s. The center-to-center distance was measured and twice this value was taken to be the wavelength λ_G.[20] The experimental values are mean values obtained from 50 to 100 samples. The reported values are best estimated for the results and, with 95% confidence, the true value is believed to lie within $\pm 2\%$ of the present value of tap water. The *solid line* in Fig. 5.13 is the experimental result for Newtonian fluids.[21] Although the experimental data for tap water lie slightly above this line, there is a tendency for greater agreement at lower Taylor numbers. The wavelength of polymer solutions decreases slowly with Re in comparison to tap water, for all concentrations. This means that the solutions' stabilizing effect on the Gőrtler instability increases with Re and with polymer concentration.

The six cells appear in the Taylor vortex flow of the stable mode for this experimental apparatus because the aspect ratio of the cylinder height and the gap width is $\Gamma = 6.15$. The experimental results also showed the

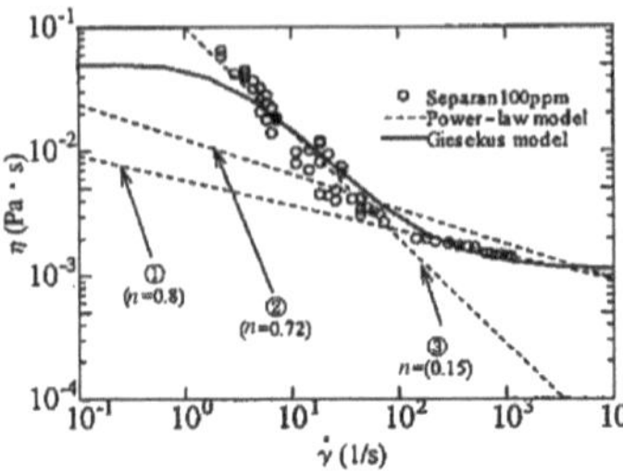

Figure 5.14 ***Apparent viscosity of Separan 100 solutions.***

phenomena. Generally speaking, the flow is analyzed by the momentum equation for laminar flow. Watanabe et al.[17] analyzed the Taylor vortex flow of the stable mode to clarify the effects of viscoelasticity on the flow behavior by using the Giesekus model.[22] The finite-element-based simulation program used was POLYFLOW (Fluent Inc.). The calculated fluid is a viscoelastic fluid, which is equal to Separan 100 ppm solution in terms of the physical constants. In addition, three cases were calculated using the power-law model to examine the effect of the non-Newtonian viscosity.

Fig. 5.14 shows the apparent viscosity of the Separan 100 ppm solution, which calculated using the Giesekus model and power-law models. The power-law model fits the experimental data for many pseudo-plastics at an intermediate shear rate and over a range of 10- to 100-fold, but fails both at very low and very high shear rates as shown in the figure. The first normal stress difference of the Separan 100 ppm solution was determined by extrapolation from the high concentrated solutions data, because the experimental data is lacking for dilute polymer solutions.

Fig. 5.15 shows the analytical results of the power-law model. In Fig. 5.15, *n* is the power law exponent, and these flow curves of the model are shown in Fig. 5.14. Increased viscosity has a stabilizing effect on the inertial instability; however, the results of the power-law model do not concur with the experimental results. The analytical results of the Giesekus model are shown in Fig. 5.16 in comparison with the experimental results of the Separan solutions. In Fig. 5.16, the *dotted lines* show the position of Taylor cell No. 1 and No. 6 of the glycerin 60% solution for the case of Newtonian fluids. Taylor cell No. 1 and No.6 are the cells that appear at the bottom and the top of the gum, respectively. The calculated results of the viscoelastic fluids agree with the experimental results.

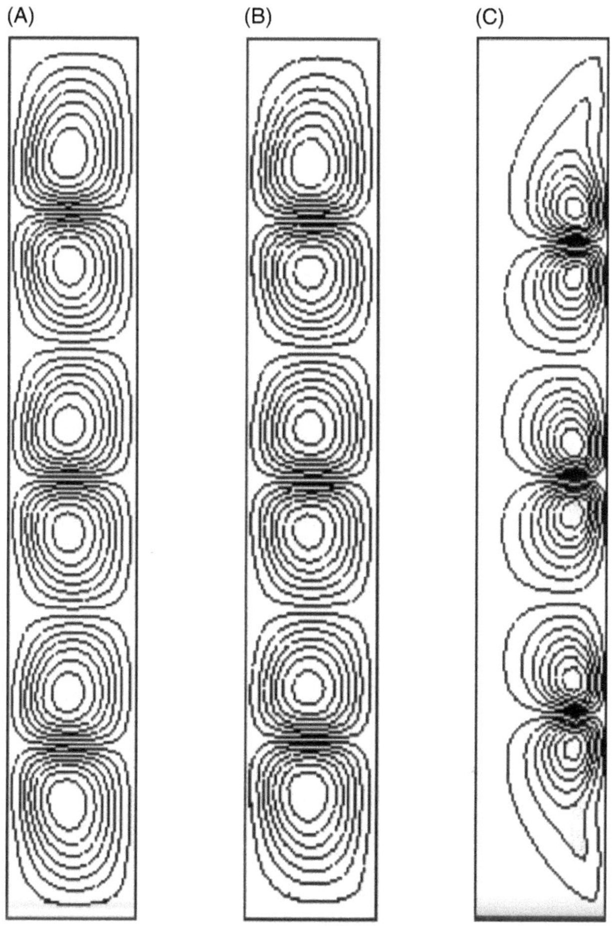

Figure 5.15 ***Analytical results for power-law models.*** (A) $n = 0.8$; $N_{Re} = 0.012$, (B) $n = 0.72$; $N_{Re} = 1.62 \times 10^{-3}$, (C) $n = 0.15$; $N_{Re} = 1.26 \times 10^{-8}$.

5.2.3 Flow Around a Rotating Disk

Flow around a rotating disk is closely related to the behavior of three dimensional boundary layer or the internal flow in an impeller of turbo-machinery. Thus, many studies have been done analytically or experimentally for it. The flow analysis is divided by two categories that a disk rotates in a still infinity fluid and a disk rotates in housing. About the former, the analytical results of frictional resistance of a Newtonian fluid have been reported by Von Kármán,[23] S. Goldstein,[24] Fujimoto and Iwagaki,[25] and Ito.[26]

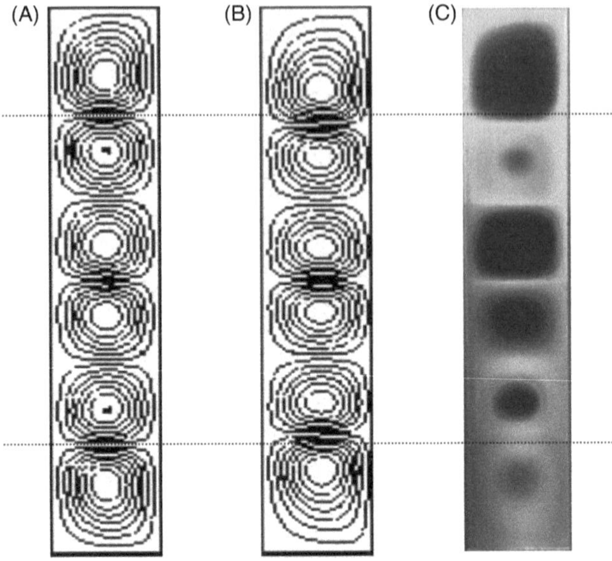

Figure 5.16 ***Comparison between analytical and experimental results.*** (A) Glycerin 60 wt% (Re = 80.8), (B) Viscoelastic model (NRe=1.62 × 10) (We = 25.1), and (C) Separan 100 ppm.

The disk is rotated at a constant angular velocity ω in a fluid of infinite extension (see Example 2.2 in Chapter 2 for the derivation of Eqs. (5.5) and (5.6)), and the boundary layer exists on the rotating disk surface. The momentum integral equations of the boundary layer on the rotating disk are given as follows:

$$\frac{d}{dr}\left(r\int_0^\delta u_r^2\,dz\right)-\int_0^\delta u_\theta^2 dz=-\frac{\tau_r r}{\rho} \tag{5.5}$$

$$\frac{d}{dr}\left(r^2\int_0^\delta u_r u_\theta\,dz\right)=-\frac{\tau_\theta r^2}{\rho} \tag{5.6}$$

where u_r and u_θ are the velocity components of the fluid in the directions of r and θ and the axial velocity component, and τ_r and τ_θ are shearing stress components at the surface. δ is the boundary layer thickness. In the case of turbulent flow, the terms of velocity profile and the stress in Eqs. (5.5) and (5.6) are used for the time mean value. And also, although Kármán delivered these equations from the momentum balance for boundary layer, we can deliver by integrating N-S equation or the equation of motion over the whole boundary layer for laminar or turbulent flow, respectively.

On the other hand, Tomita[27] deduced the moment coefficient of power-law fluid using Eqs. (5.5) and (5.6). For viscoelastic fluid, Mochimaru and Tomita[28] analyzed the moment coefficient. Although we can predict the drag reduction of non-Newtonian fluids for the moment coefficient of a rotating disk from the previous works. There are few experimental studies[29] because it is necessary a large tank and a device for measuring micro torque in the experimental apparatus of this case.

An example of the experimental apparatus[30] shows in Fig. 5.17A and B. It is about 1700 L in the liquid volume. Torque acting on a rotating shaft is measured by the torque pick-up attached to the end of shaft. In the detail drawing of Fig. 5.17B, four phosphor bronze plates of 0.4 mm in thickness are soldered with the boss made of brass. The strain gauges is stuck on those plates, and waterproofing, the gap is filled with insulator material and then the cylindrical part is wrapped with the insulator tape. The torque pick-up was calibrated by applying a known static torque to the shaft. The relation between the torque and the strain in three torque pick-up shafts built on a trial gave essentially straight lines and a high accuracy of the measurement was obtained.

Fig. 5.18 shows the experimental results of moment coefficients for tap water. The *straight lines* 1 and 2 in Fig. 5.18 are Kármán' s laminar and turbulent theoretical results, respectively. It is clearly seen that the agreement between the theoretical results and the experimental data is excellent in those flow ranges. The numerical value of Reynolds number at which the transition occurs is obtained as being approximately $R_{ec} = 2.8 \times 10^5$. This value agrees with the result of Gregory et al.,[31] which were calculated using the transition position found at the number of rotational speed with the china-cray technique.

Frictional moment in the transitional range for Newtonian fluids[24] is calculated using the following equation if the transition radius r_c that is, the transitional Reynolds number R_{ec} is clear

$$M = 2\pi\left[\int_0^{r_c} r^2\left(-\tau_\theta\right)dr + \left\{\int_0^{a} r^2\left(-\tau_\theta'\right)dr - \int_0^{r_c} r^2\left(-\tau_\theta'\right)dr\right\}\right] \quad (5.7)$$

where, τ_θ and τ_θ' are the disk wall shear stress for laminar and turbulent boundary layer on a rotating disk, respectively. The solid curve in Fig. 5.13 is the calculated result obtained by substituting Rec $= 2.8 \times 10^5$ for Eq. (5.7). Experimental results fit with it.

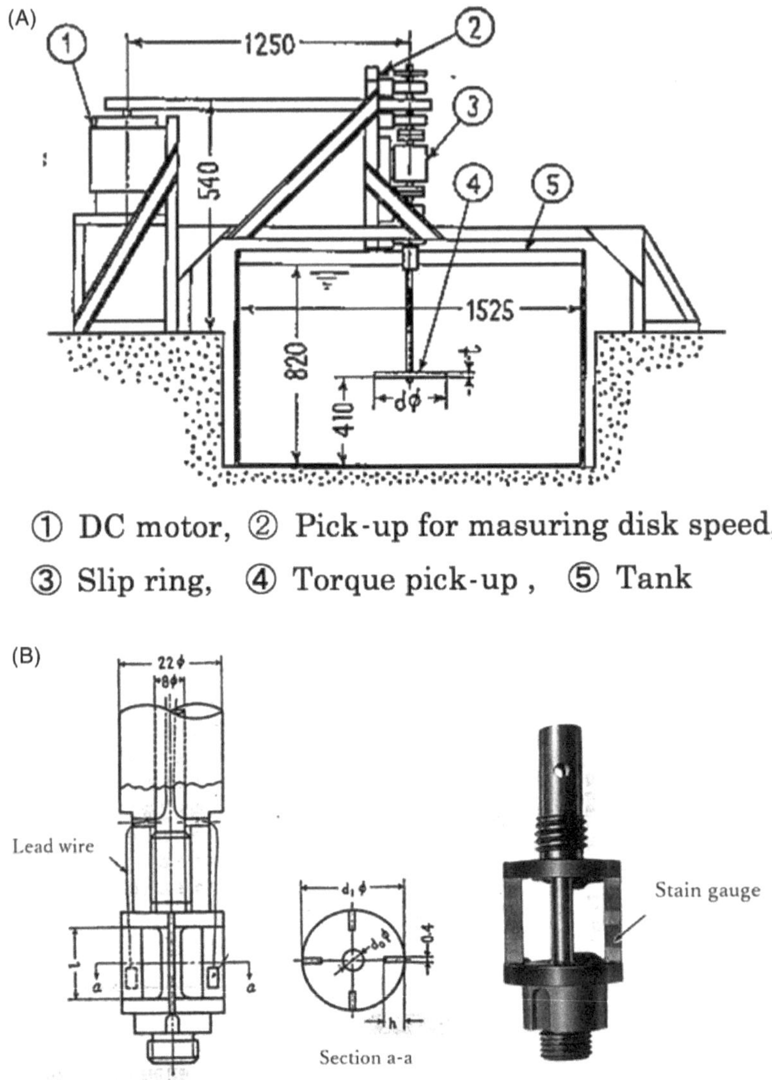

	l	d_0	d_1	h
①	14	4	22	3.5
②	14	4	25	3.5
③	14	4	28	3.5

Figure 5.17 ***Experimental apparatus.*** (A) Overall view and (B) torque pick-up detail.

Moment coefficients of dilute polymer solutions are shown in Fig. 5.19A and B. According to the increase in the concentration of polymer, the transition behavior of the moment coefficients is gently changing

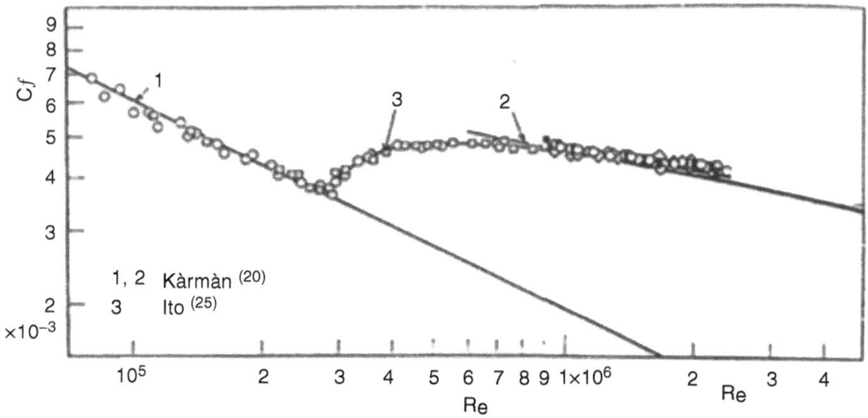

Figure 5.18 ***Moment coefficient of Newtonian fluids.***

although the experimental data increase slightly in comparison with the Kàrmàn 's formula for laminar flow. By the comparison between Separan and Peo solutions, it is seen that the quality of being instabile occurs in low concentration range for Separan solutions. This fact means that it depends on viscoelastic effect. According to the increase of the concentration of polymer, it cannot be ignored for non-Newtonian viscosity. Fig. 5.20A–C shows the experimental results for high concentration range. In the figure, the power-law index, n is defined by the power-law fluids model, $\tau^{n} = K\dot{\gamma}$. From Fig. 5.20A, it is clearly seen that Tomita's formula for power-law fluid agree well with the experimental results for HEC solutions having weak viscoelastic effect, and the data on 500 ppm solutions in the transition region are significantly lower than those on Newtonian fluids, and the theoretical results of Eq. (5.7) calculated as $R_{ec} = 1.7 \times 10^{5}$ is in good agreement with the experimental data. On the other hand, the experimental data on Peo 0.3% solutions in Fig. 5.20B show that, although the data in the range of Reynolds number smaller than 1.4×10^{5} agree with Tomita's formula, the data deviate slightly upward with an increase of Reynolds number, and the variation of the experimental data in the transition region becomes gentle, compared with the data on HEC 500 ppm solutions. In case of Separan solutions in Fig. 5.20C, a decreasing tendency of the data is seen in laminar flow region compared with Tomita's formula. Because the viscoelastic effect on the flow is not considered in the analysis, it is required for comparison between the formula and the experimental data.

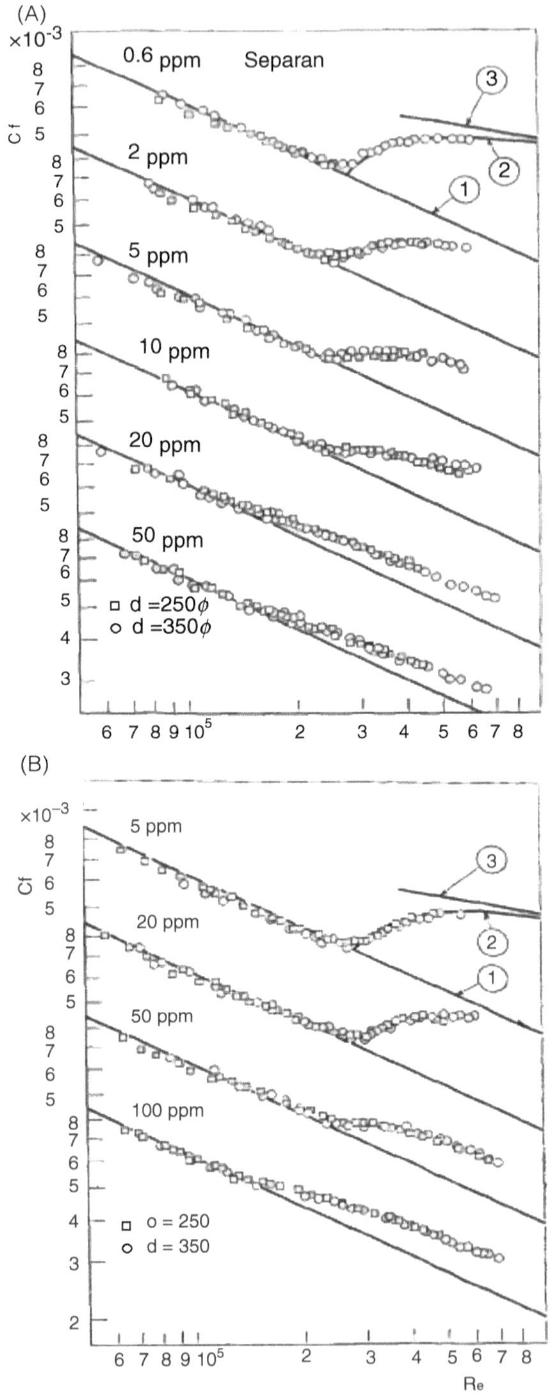

Figure 5.19 ***Moment coefficients of dilute polymer solutions.*** (A) Separan solutions and (B) Peo solutions.

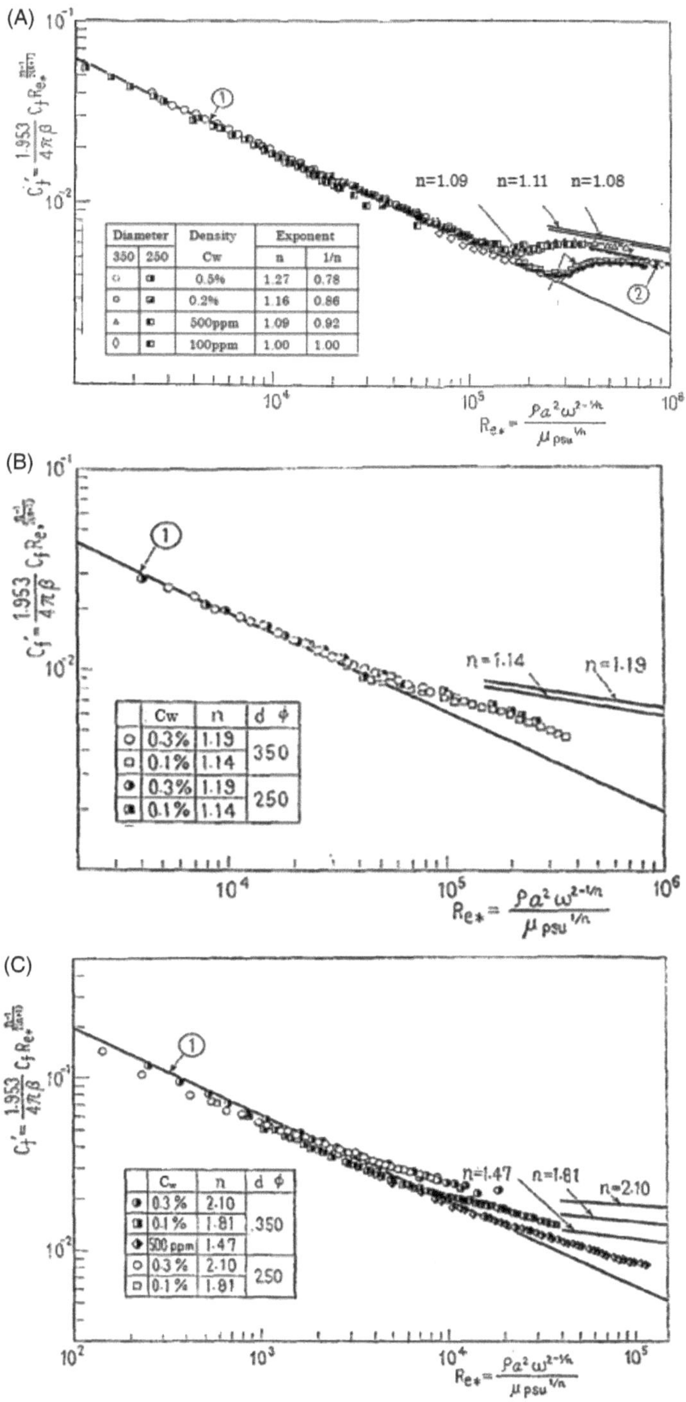

Figure 5.20 ***Moment coefficients of high concentration polymer solutions.*** (A) HEC solutions, (B) Peo solutions, and (C) Separan solutions.

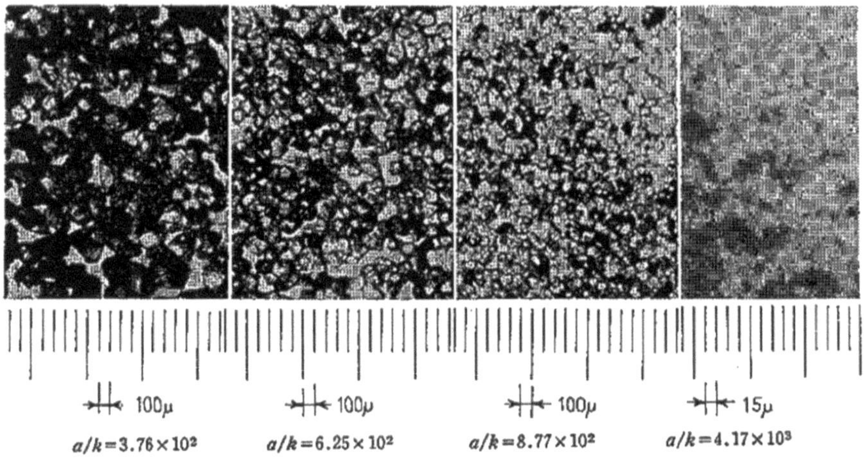

Figure 5.21 ***Micrographs of rough surface disk.***

Also, when the surface roughness of disk increased, the moment coefficient does not depend on Reynolds number. It may be significant to clarify experimentally the effect of the surface roughness of a disk on the moment coefficient because the data is insufficient at present. Fig. 5.21 shows the micrographs of rough disk surface.[30] The rough disks with the diameter of 170 mm on the two faces of which the sand-grains of various sizes were cemented, were used. The mean size of the grains was decided as the mean values of the sieve's mesh. Also, these values were confirmed by the measurement for the distributions of the particle size.

In each of the rough disks, the value of moment coefficient holds a constant value with an increase of Reynolds number as shown in Fig. 5.22. Experiments show that the complete rough flow was obtained. The analytical results of Fujimoto and Iwagaki are indicated in Fig. 5.22, so as to be compared with the experimental data. As the value of relative roughness, (a/k) that it is a ratio the disk radius and the protrusion height decreases, the difference between the analytical and experimental values increases further. However, as pointed out by Ito it may be concluded that the formula agrees well with the experimental data in the range of the larger value of (a/k).

The experimental results of dilute polymer solutions are shown in Fig. 5.23(a) and (b). As is evident from Fig. 5.23A where the cemented grains are fly ash, depend more upon Reynolds number according to the increase of the concentration. From this fact it is considered that disk's surface is the state of hydraulic smooth surface. The value of drag reduction ratio increases with an increase of (a/k), and the trend increases further

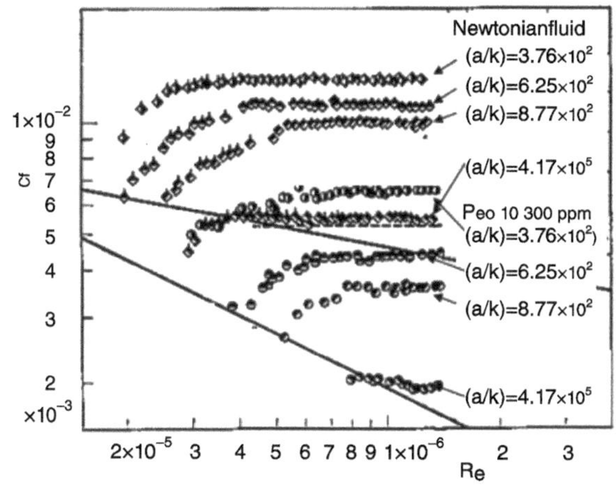

Figure 5.22 ***Moment coefficients of a rough surface disk in Newtonian fluids.***

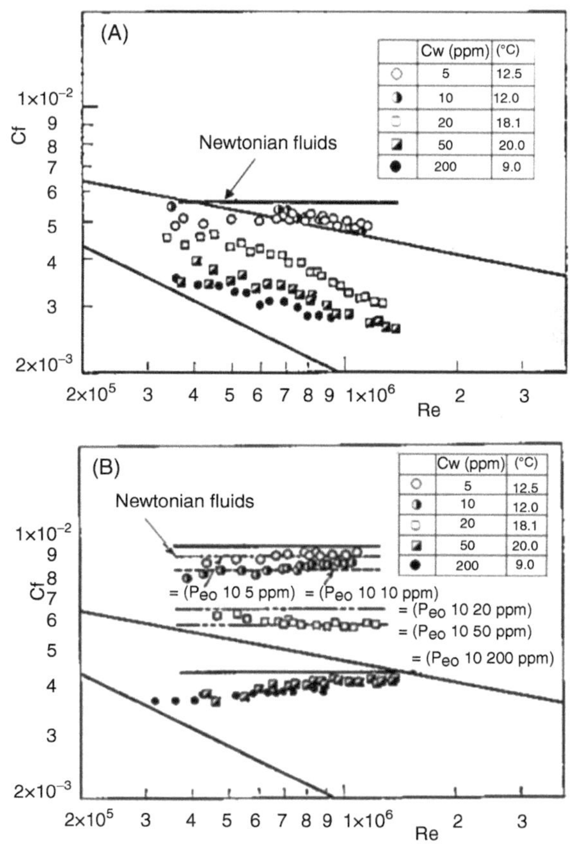

Figure 5.23 ***Moment coefficients of a rough surface disk in Peo 10 solutions.*** (A) $a/k = 4.17 \times 10^{-3}$ and (B) $a/k = 3.45 \times 10^{-3}$.

with an increase of the concentration of the solutions. A conclusion cannot be drawn due to an insufficient number of the experimental data, but the relation between the thickness of viscous layer and the viscoelastic property may be analogized with these experimental data.

It will be necessary to research the characteristics of boundary layer on a rotating disk by a flow visualization technique to grasp the drag reduction phenomena on the moment coefficient. The flow pattern[31] in the boundary layer was visualized by a oil film technique, and the deference of flow angle[32] between Newtonian fluids and dilute polymer solutions was clarified.

The flow angle ϕ is defined as follows,

$$\tan\phi = -\frac{\left(\partial v_r/\partial z\right)_{z=0}}{\left(\partial v_\theta/\partial z\right)_{z=0}} = -\frac{\tau_{rz|z=0}}{\tau_{\theta z|z=0}} \tag{5.8}$$

Kármán gives $\phi = 39.6°$ for laminar flow from the analytical result of the velocity profile. Also in the case of turbulent flow, if ϕ coincides with the direction of the flow near the surface of the disk, it is supposed that shows the ratio of the wall shearing stresses in the same manner as the laminar flow.

Fig. 5.24A and B shows the streaks of Newtonian fluids and the polymer solutions obtained by an oil film technique, respectively. The thickness of the coated oil film is about 50 μm and it is smaller than the boundary layer thickness of the disk. It is seen that the Reynolds number of Peo 20 ppm

(A) (B)

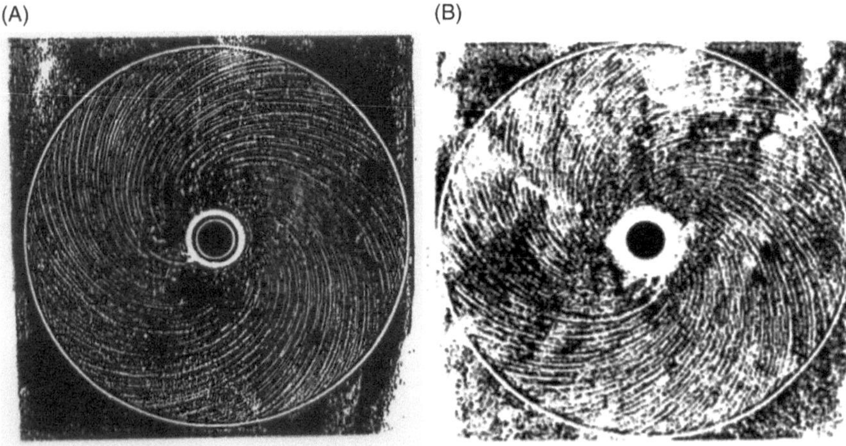

Figure 5.24 ***Streaks of flow on a rotating surface.*** (A) Newtonian fluids (Re = 1.55 × 10^6) and (B) PEO 18 (Re = 2.21 × 10^6).

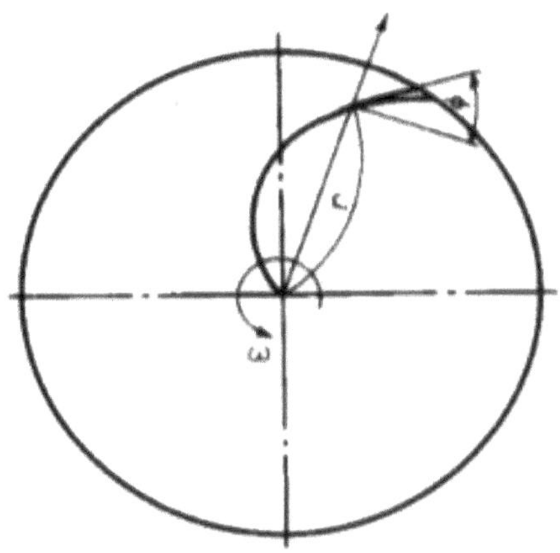

Figure 5.25 ***A path line of a rotating flow.***

solution is larger than that of tap water but the streak is not positioned closer to the center.

By referring the definition for a path line on a rotating disk as shown in Fig. 5.25, the experimental data of flow angle[33] is obtained from Fig. 5.24. Figs. 5.26 and 5.27 show the result of a rotating disk in large tank and an enclosed disk, respectively. It is seen from Fig. 5.26A that the experimental value of ϕ in Newtonian fluids is slightly larger than Cochran's exact solution, $\phi = 39.6°$ in the range of R_{er} less than 3×10^5, and the average values are $\phi = 44.1 \pm 2.5°$. In the turbulent flow range, the experimental data are hardly scattered, and they are arranged in a straight line as shown in the figure. The experimental value of the dilute polymer solutions in the laminar flow range in Fig. 5.26B, is $\phi = 43.5 \pm 2.5°$ and $\phi = 41.1 \pm 2.5°$ in Separan 20 ppm and 100 ppm solutions, respectively, and the experimental value decreases with an increase in the concentration of the solutions. Thereby it can be considered that the scattering of the experimental value $\pm 2.5°$ is the probable result of this experiment on the oil film technique having been done in liquid flow, and the range measurement being the neighborhood of the transition region. In turbulent flow range, the change of the experimental value with Reynolds number is gentle compared with value of Newtonian fluids, and the value of ϕ increases with an increase in the concentration of the solutions, which is contrary to the experimental result in the laminar flow range.

For the enclosed rotating disk, it is seen from Fig. 5.27A that the value of Newtonian fluids almost coincides with the exact solution within the low

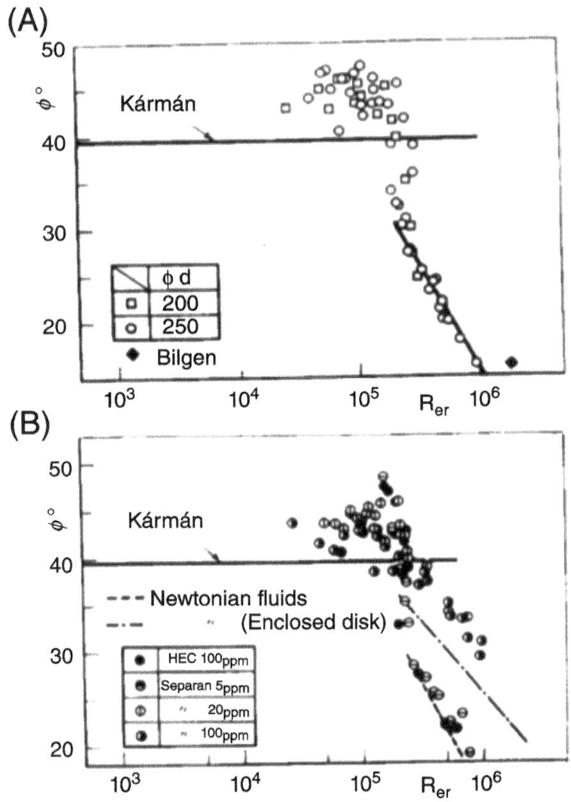

Figure 5.26 ***Flow angles of a rotating disk in a large tank.*** (A) Newtonian fluids and (B) dilute polymer solutions.

Reynolds number range, but the data tend to slightly increase with an increase in Reynolds number and they scatter in the range of $R_{er} = 2\sim3 \times 10^5$. The result which gives essentially a straight line in the turbulent flow range is like the case of the disk in a large tank. For a given Reynolds number in the turbulent flow range, the value of ϕ in the enclosed disk is larger than that in the disk in large tank. It had been seen from the result of the frictional torque measurement that the moment coefficient C_f in the range of (s/d) that it is a ratio the gap and disk diameter used in this experiment well agree with Kármán' s formula within the laminar flow range, and as the value of (s/d) increases, C_f increases in the turbulent flow range, But the experimental value of ϕ in the turbulent flow range in Fig. 5.27A does not clearly show the effect of the variation in (s/d).

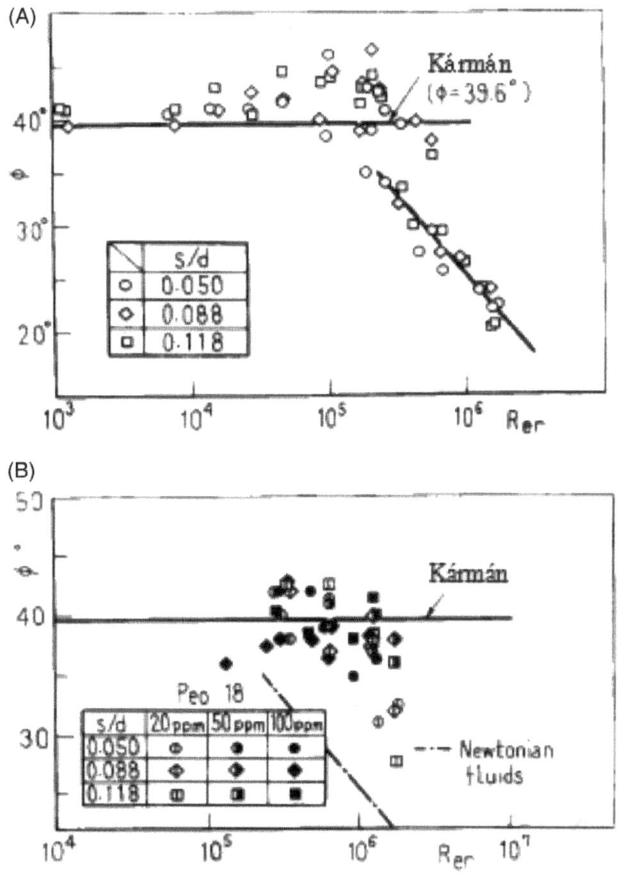

Figure 5.27 ***Flow angles of an enclosed rotating disk.*** (A) Newtonian fluids and (B) dilute polymer solutions.

For Peo 18 20 ppm solutions in Fig. 5.27B, ϕ decreases almost parallel with the data of Newtonian fluids for R_{er} larger than 8×10^5, and ϕ of 100 ppm solutions is slightly smaller than that in the exact solution, and does not change appreciably with an increase in Reynolds number. These experimental results in the laminar flow range approximately agree with the results of the disk in large tank. As the non-Newtonian viscosity effect in the dilute polymer solutions can be neglected, the difference of the flow condition on the surface of the rotating disk between Newtonian fluids and the dilute polymer solutions is due to the viscoelastic effect, and a decrease of ϕ means an increase of the tangential component of the disk wall shear stress.

5.2.4 Flow Past a Sphere

There are some experimental studies about the drag reduction in polymer solutions. Ruszczycky[34] showed that the maximum drag coefficient reduction is 28% in 0.75% polyethylene oxide. White[35] carried out tests with polyethylene oxide aqueous solutions of 0.005, 0.0075, 0.01, and 0.0015% concentrations. The maximum drag coefficient reduction was 45% in the 0.0075% solutions at $Re = 3.8 \times 10^4$. Kato et al.[36] carried out experiments using dilute Peo solutions of 0.002~0.5% concentrations. It was found that maximum reductions occurred in Peo 10, Peo 16, and Peo 18 solutions of 0.015, 0.0015, and 0.005% concentrations, respectively. The maximum drag coefficient reductions are 20%~40%. And they clarified the move of separation point and the decrease of width of a wake occur in 150 ppm Peo solution in Reynolds number range $Re > 10^4$. White[37] carried out tests for range of Reynolds number to span the critical region using dilute polyethylene oxide WSR 301 aqueous solutions. The drag coefficient in the range below the critical Reynolds number agreed with those of previous investigations. In the critical Reynolds number range, increasing the concentration of polymer solutions increases the drag. However, there are few studies about the drag of a sphere in dilute polymer solutions in high Reynolds number range.

On the other hand, it is an important thing to clarify the onset of the drag reduction phenomenon for flow around a sphere to consider the analogy for the flow of fine solid particle suspensions. Fig. 5.28 shows the onset[38] of the drag reduction phenomenon for the drag coefficient of Separan solutions. In the figures, the *solid*[39] and the *dotted lines*[40] represent the experimental result for Newtonian fluids. The drag reduction occurs at Newton range where the drag coefficient is almost a constant value. Experimental data in Fig. 5.28 show that the onset Reynolds numbers are $Re = 5 \times 10^3$ and $Re = 1.3 \times 10^4$ in Separan solutions 100 ppm and 50 ppm, respectively. As described in Chapter 4, the boundary layer is laminar flow and it changes to turbulent flow by slight disturbance. It was drawn on an analogy between the flows of fine solid particle suspensions and dilute polymer solutions for the drag reduction phenomenon, in which a viscoelastic property is present. Boundary layer of a sphere becomes a turbulent flow with an increase of Reynolds number and the drag coefficient decreases sharply at the critical Reynolds number range. Experimental data on the fall sphere test in near flow range are shown in Fig. 5.29A and B. L_s in those graphs is the fall

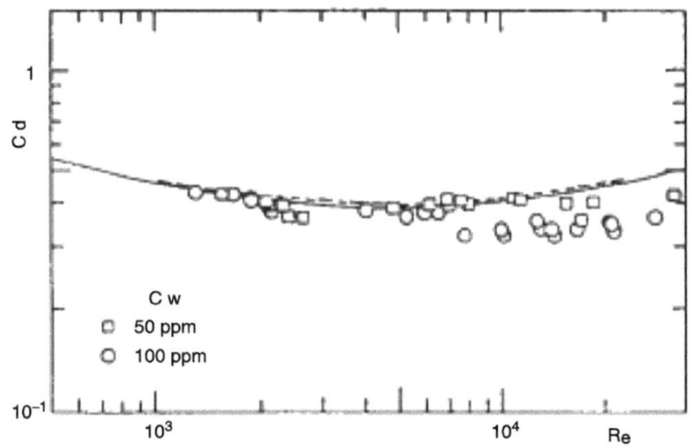

Figure 5.28 ***Onset of drag reduction of Separan solutions.***

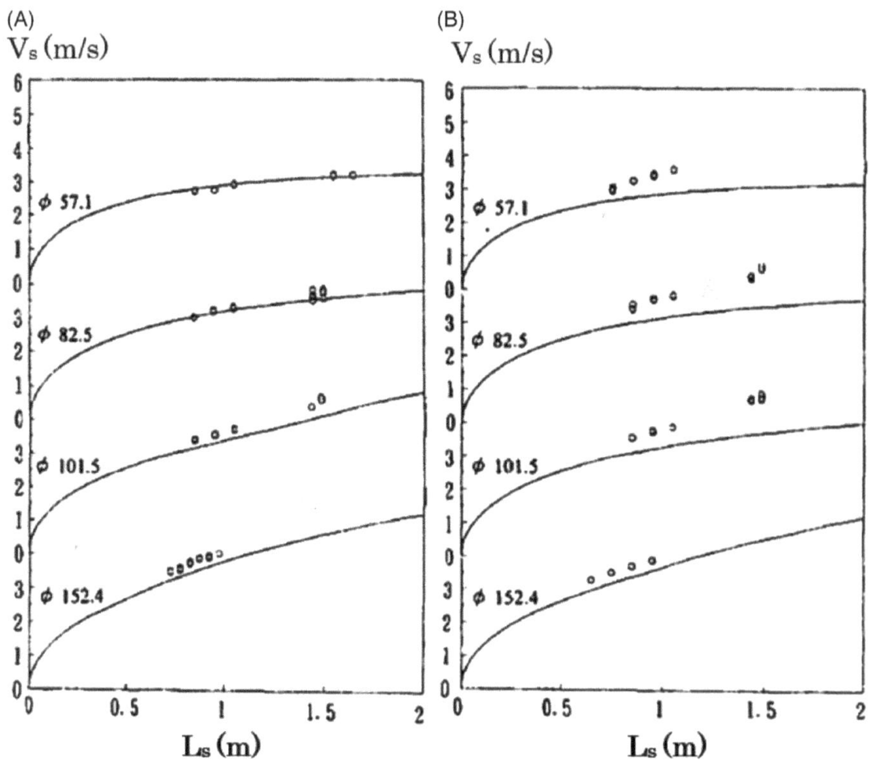

Figure 5.29 ***Fall velocity.*** (A) Tap water and (B) Separan solutions.

length from water surface in the tank, and the *solid lines* are the calculated result for Eq. (4.14) in Chapter 4. Fig. 4.29A shows that the experimental data for the sphere with d_s = 57.1 mm lay on the solid line, but those for d_s = 82.5~152.4 mm lay slightly above the solid line. The possible causes of these discrepancies are the effect of (1) the water surface, (2) the bottom of the tank, and (3) the wall of the tank. Regarding (1), through experiments conducted by changing the distance between the stopper and the water surface, it was proven that there is no effect of the water surface on the results. The distance for (2) is more than 3 sphere diameters in this study. Although the effect of the tank wall in the small Reynolds number range has been investigated by many researchers, there are very few studies of this effect in the large Reynolds number range. Generally, if the motion of a sphere is affected by the wall of the tank during its fall in the small Reynolds number range, the fall velocity will decrease compared to the case in which there is no wall effect.

Fig. 5.30 shows the relationship between the fall velocity ration (V_{se}/V_{sc}) and the diameter ratio (d_s/D). V_{se} and V_{sc} are an experimental value of the fall velocity of the sphere and the corresponding calculated value, respectively. The diameter ratio is a ratio of the sphere diameter and tank diameter. L_t is the distance from the starting point to the bottom of the test tank. Experimental data show that the fall velocity increases, compared to that of the calculated result, with the increase in the diameter ratio. However, beyond a certain value of the diameter ratio, the fall velocity ratio becomes constant even with further increase of the diameter ratio. The uncertainties in the experimental results of the fall velocity are given that the reported values are the best estimated of the results, and with 95% confidence, the

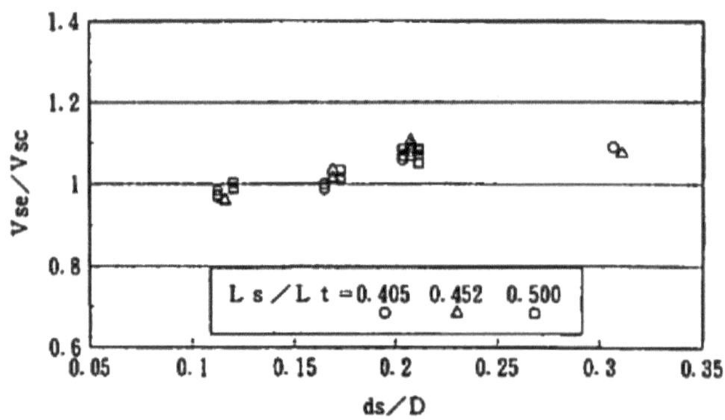

Figure 5.30 ***Relationship between fall velocity ratio and diameter ratio in tap water.***

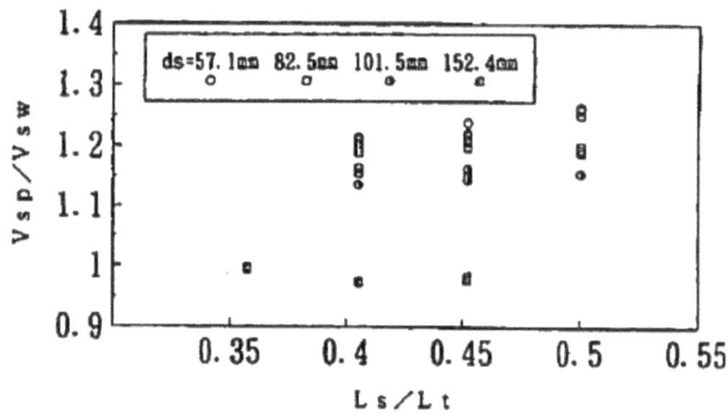

Figure 5.31 ***Ratio of velocities in Separan solutions 20 ppm to that in tap water.***

true value is believed to lie within ±3.74 % of the present value for the sphere with d_s = 101.5 mm in tap water.

Experimental data of Separan solutions are larger than calculated values for the four spheres as shown in Fig. 5.29B that the *solid lines* show the calculated results obtained by using the kinematic viscosity of solution. Fig. 5.31 shows an example of the ratio of fall velocity of Separan solutions to that of tap water. In this figure, V_{sp} represents the fall velocity of the sphere in solutions and V_{sw} represents that of tap water. The maximum increase in the fall velocity ratio is about 26%.

In general, the flow phenomenon around a sphere becomes easily understandable from the graph on the relationship between the drag coefficient and Reynolds number. Fig. 5.32 shows the drag coefficient of a sphere rearranged from the experimental data on the fall velocity using Eq. (4.19) in Chapter 4. Separan solutions 5 ppm shows a tendency similar to tap water. On the other hand, Separan solutions 20 ppm and 50 ppm show drag reduction in range the $Re \leq 3.0 \times 10^5$ range. However, this reduction does not occur above the critical Reynolds number. In other words, the polymer addition does not lead to a rapid change at the critical Reynolds number range.

Since polymer solutions have a viscoelastic property as shown Fig. 4.40 in Chapter 4, the drag coefficient of a sphere is given by a dimensional analysis as follows:

$$Cd = f(\mathrm{Re}, We) \tag{5.9}$$

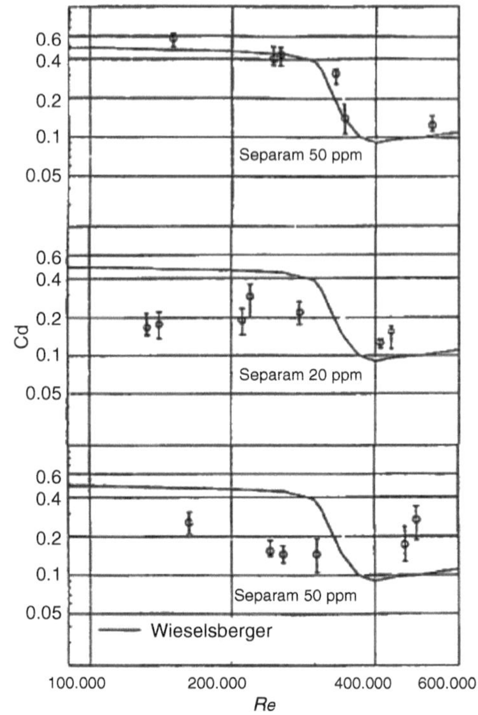

Figure 5.32 ***Drag coefficients of sphere in high Reynolds number range.***

where, *We* is Weissenberg number and it is given from Eq. (4.19) in Chapter 4 using the condition of dilute solution.

$$We = \frac{\lambda}{\mu}\left(\frac{V}{d}\right)$$

The characteristic length is the diameter of a sphere *d* and the characteristics velocity is the fall velocity of a sphere *V*.

Fig. 5.33 shows the relationship between the drag reduction ratio and the Weissenberg number. Like Eq. (4.21) in Chapter 4, the drag reduction ratio is defined as $(\Delta C_d/C_{dw}) = \left(C_{dw} - C_d\right)/C_d$, where C_{dw} and C_d are the drag coefficient of tap water and the polymer solutions, respectively. For the Weissenberg number of 0.5, that is, for a 5 ppm solution, the drag reduction scarcely occurs. At the Weissenberg number rang $We = 3{\sim}10$ that is, for 20 and 50 ppm solution, the drag reduction ratio scarcely occurs, and the drag reduction ratio is 50%–70% then. Although we can rearrange the drag reduction ratio and the range by this idea, much remains to be done, both in terms of the application of the rheological equation and the measurement

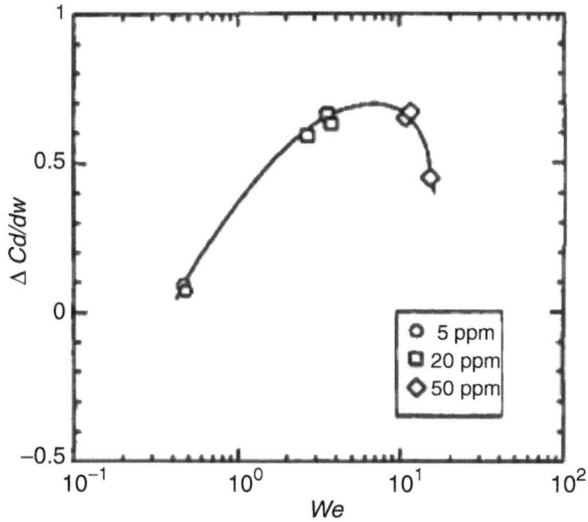

Figure 5.33 ***Relationship between drag reduction ratio and Weisenberg number.***

for the viscoelastic physical constant. It will be also necessary to approach this from the viewpoint of extensional flow properties of dilute polymer solutions in future.

The causes of drag reduction on flow around a sphere may be explicitly predicted by the flow visualization result. Kato et al.[36] clarified the flow around a sphere in dilute polymer solutions by the flow visualization as shown in Fig. 5.34A–D In this figure, A and B indicate the result for Newtonian fluids and 0.015wt% Peo solutions, respectively. The flow was visualized by a fall sphere that the surface was coated with water-soluble paint. From Fig. 5.34A and B, the difference in the wake of a sphere in not recognized. However, it is clear and obvious from Fig. 5.34C and D that Reynolds number being over 10^4, occurs a remarkable difference in two flows. It is seen that the separation point of polymer solutions moves to the downstream side in comparison with that of tap water in the range of Re $\geqq 1.3 \times 10^4$.

The flow in the range is an unstable laminar boundary layer and the form drag plays a dominant role for the total drag of a sphere. As a result, it can be considered that the drag reduction occurs in the Reynolds number range.

5.3 SURFACTANT SOLUTIONS

5.3.1 Flow in a Pipe

Because the degradation does not occur in the flow of surfactant solutions in the pipe line systems with a centrifugal pump, the technique has been applied to the heat medium transport system. However, the flow

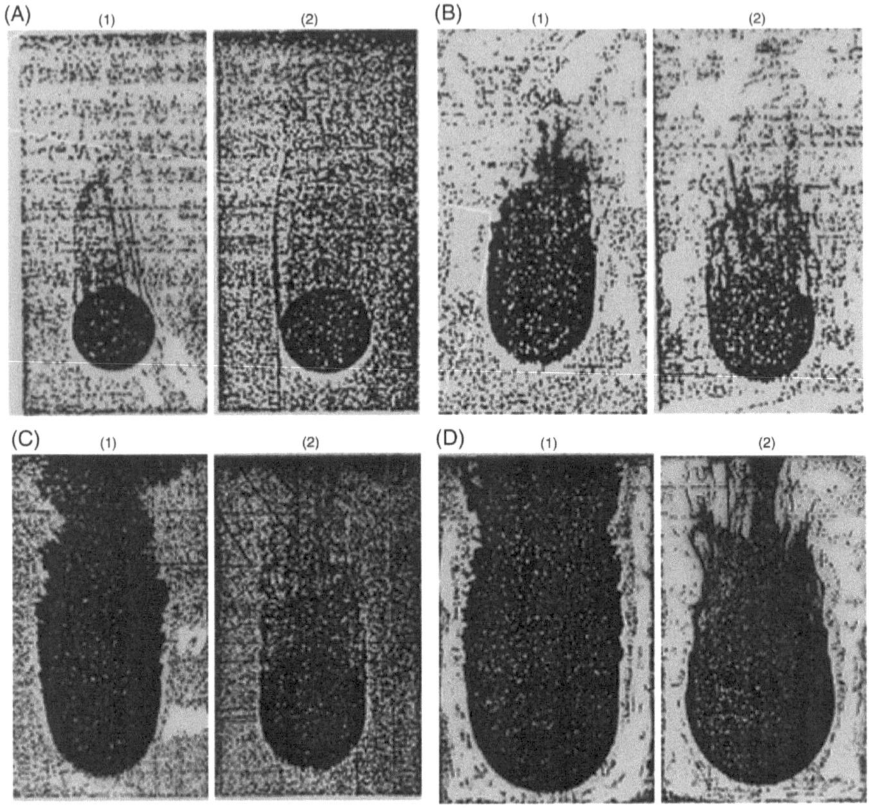

Figure 5.34 ***Flow visualization results for flow around a fall sphere.*** (A) Re = 1.3×10^3, (B) Re = 6.0×10^3, (C) Re = 1.3×10^4, and (D) Re = 2.2×10^4.

characteristics are complex and an adequate rheological formula is yet to be established. Regarding the mechanism of the drag reduction effects of surfactant aqueous solutions, micelles—aggregations of surfactant molecules—are known to form networks that inhibit turbulence. Because few studies have been conducted on the phenomena associated with the formation of micelles, many aspects of drag reduction and the formation of micelles remain unknown. In addition, it is important to clarify the effects of various additives such as rust inhibitors on the drag-reducing capacity of surfactant aqueous solutions for practical applications. With respect to additives, Haruki and Horibe[41] investigated the effect of propylene glycol and ethylene glycol on the drag reduction and the rheological properties of oleyl dihydroxyethyl amine oxide (ODEAO). They reported that the drag reduction Reynolds number range expanded compared to that of water. Zhang et al.[42] reported that ethylene glycol has a significant effect on the drag reduction and the rheological properties of oleyl bis (hydroxyethyl) methylammonium chloride (Ethoquad O/12) and oleyltris (hydroxyethyl) methylammonium acetate (Ethoquad O/13). They reported that the drag reduction percentage decreases because the formation of the micelles is disrupted by the addition of ethylene glycol. Incidentally, not much data exists regarding the effects of various additives on the formation of micelles.

In this section, the effects of additives such as rust inhibitors, propylene glycol, coconut fatty acid, and potassium hydroxide on the formation of the micelles and the drag reduction capacity of a surfactant aqueous solution, namely, Ethoquad O/12, by measuring the particle size distributions using a dynamic light-scattering spectrophotometer and the pressure loss of the pipe are described.

The test fluids used in this study[43,44] are listed in Table 5.1. NDT-1 is an aqueous solution containing only coconut fatty acid and potassium hydroxide. NDT-2 is a surfactant aqueous solution of oleyl bis-hydroxy-ethyl ammonium chloride (Ethoquad O/12, a cationic surfactant), to which sodium salicylate is added. NDT-3 contained coconut fatty acid and potassium hydroxide in addition to the components of NDT-2. For comparison, experiments are also performed using an aqueous solution (NDT-4) of propylene glycol, which is generally used as an antifreeze for water-based coolants, and a solution (NDT-5) produced by adding propylene glycol to NDT-2. Regarding rust inhibitors, NDT-6, NDT-7, and NDT-8 are derived from the aqueous solution NDT-5 by respectively adding benzotriazole, amine salts, and tolyltriazole. NDT-9 and NDT-10 are decyl glucoside and lauryl glucoside aqueous solution, respectively. These alkylpolyglycoside aqueous solutions have drag reduction characteristics that are different from those of Ethoquad aqueous solutions. Such nonionic surfactant solutions have attracted interest in recent times owing to the low environmental impact of their drag reduction agents. Unfortunately, however, much higher surfactant concentrations are required to achieve the drag reduction.

The viscosities of the test fluids were measured using a capillary viscometer. As is well known, Ethoquad O/12 has a non-Newtonian viscosity. Fig. 5.35 shows the ratio of the viscosity of the test fluids to that of NDT-2 with respect to the shear rate. In the figure, η and $\eta_{NDT\text{-}2}$ are the viscosities of the test fluid and NDT-2, respectively. Only the viscosity of NDT-3 is lower than that of NDT-2, and that of NDT-1 is the same as that of NDT-2. The relative highness of the viscosity of NDT-5 increased with decreasing shear rates. However, the addition of the rust inhibitors to NDT-2 increased the resultant viscosities of NDT-6, NDT-7, and NDT-8.

The particle size distributions are shown in Fig. 5.36A–C. They are measured by using a dynamic light scattering spectrophotometer. The horizontal

Table 5.1 Test solutions

Mass %	NDT-1	NDT-2	NDT-3	NDT-4	NDT-5	NDT-6	NDT-7	NDT-8	NDT-9	NDT-10
Distillated water	98.68	99.75	98.43	90.00	89.75	89.75	89.75	89.75	98.75	98.75
Ethoquad O/12	-	0.125	0.125	-	0.125	0.125	0.125	0.125	-	-
Sodium salicylate	-	0.125	0.125	-	0.125	0.125	0.125	0.125	-	-
Coconut fatty acid	1.00	-	1.00	-	-	-	-	-	-	-
Potassium hydroxide	0.32	-	0.32	-	-	-	-	-	-	-
Decyl glucoside	-	-	-	-	-	-	-	-	1.25	-
Lauryl glucoside	-	-	-	-	-	-	-	-	-	1.25
Propylene glycol	-	-	-	10.00	10.00	10.00	10.00	10.00	-	-
Benzotriazole	-	-	-	-	-	0.1	-	-	-	-
Amine salt	-	-	-	-	-	-	0.1	-	-	-
Tolyl triazole	-	-	-	-	-	-	-	0.1	-	-
Temperature °C	26.5	26.5	27.5	27.5	27.5	17.0	19.0	19.0	12.0	12.0

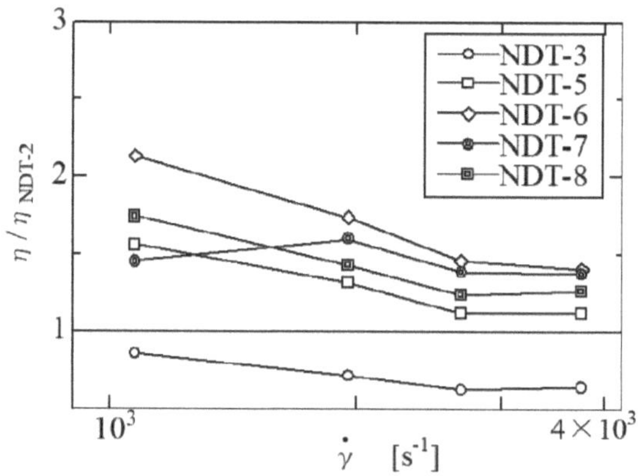

Figure 5.35 ***Effect of additives on viscosity of Ethoquard O/12.***

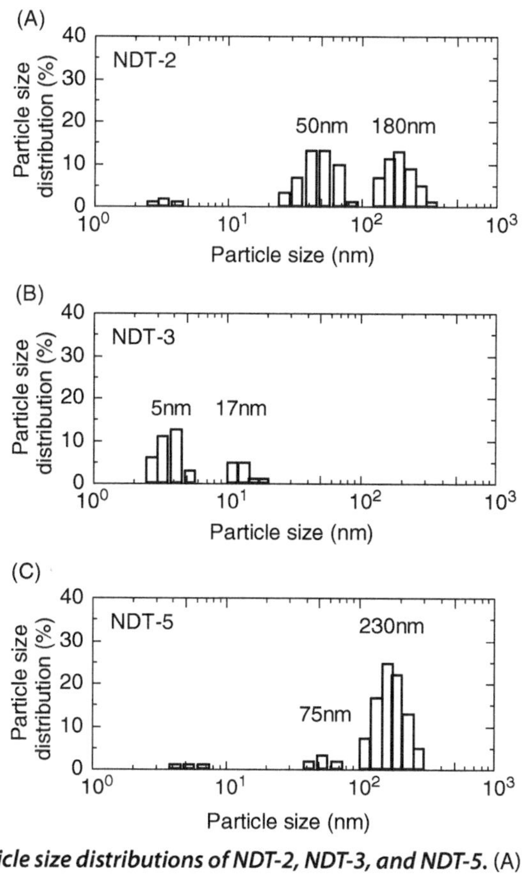

Figure 5.36 ***Particle size distributions of NDT-2, NDT-3, and NDT-5.*** (A) NDT-2, (B) NDT-3, and (C) NDT-5.

and vertical axes respectively represent the particle size and percentage of particles of a particular size with respect to the total number of particles. NDT-2 was observed to have two peaks near 50 and 180 nm, respectively, as shown in Fig. 5.36A. The existence of particles was not confirmed in the solution produced by simply adding the Ethoquad O/12 aqueous solution to distilled water, and in that produced by adding only salicylic acid to distilled water. The particle size distributions of these solutions could therefore not be determined. The particle size distributions of NDT-1 and NDT-4 were likewise indeterminable. Furthermore, information about the aspect ratio of the rod-like micelles could not be directly obtained from the particle size distributions determined by the dynamic light-scattering spectrophotometer, although the results indicated the existence of two types of particles (50 and 180 nm in size) within the solution that were considered to be micelles. Although no direct comparison could be made between the two types of particles, owing to differences between the surfactants, their dimensions roughly corresponded to those of rod-like micelles in the surfactant aqueous solution reported by Haruki and Horibe. These two peaks were therefore considered to be the sizes of the rod-like micelles formed by the surfactant, and particles of such sizes are considered to significantly contribute to drag reduction.

It is well known that the drag reduction capacity of a surfactant solution is not neutralized by mechanical shear such as that produced by a pump. To investigate the effect of the pump operating time on the particle size distributions, the pressure drop and particle size distributions for NDT-2 were measured simultaneously. It was observed that the pump operating time had minimal effect on the particle size distributions, and we therefore conjectured that continuous operation of the pump similarly had little effect on micelle formation in the NDT-2 solution. It should be noted that the experimental apparatus comprised a stainless steel centrifugal pump with a 25 mm bore, a stainless-steel pipe with a 15 mm inner diameter, and an electromagnetic flow meter. The experiment was performed using a pump speed of 1450 rpm and Re = 72,000. The drag reduction rate did not change perceptibly, even after 8 h of pump operation, and the degradation phenomenon was also absent.

The particle size distribution for NDT-3, in which coconut fatty acid and potassium hydroxide were added to the Ethoquad aqueous solution, is shown in Fig. 5.36B. The figure shows two peaks of the particle size distribution near 5 and 17 nm respectively, which are extremely small compared to those of NDT-2. There were hardly any rod-like micelles in NDT-3

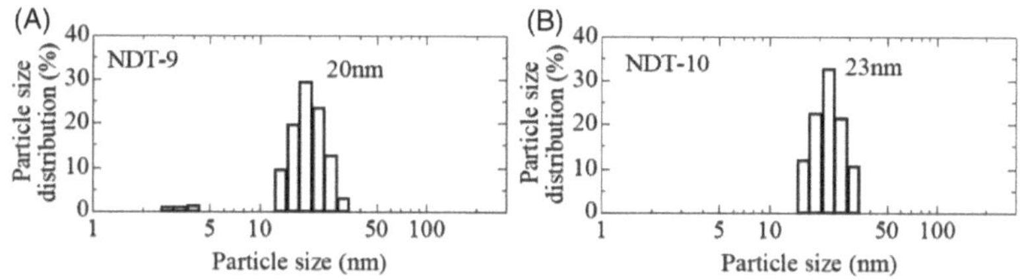

Figure 5.37 ***Particle size distributions of NDT-9 and NDT-10.*** (A) NDT-9 and (B) NDT-10.

owing to the small sizes of the particles compared to those of the micelles described earlier.

The results for NDT-5 are shown in Fig. 5.36C. As can be seen from the figure, there were two peaks in the particle size distribution near 75 and 230 nm, respectively, which are significantly larger than those of NDT-2. Assuming no change in the shape of the rod-like micelles, this suggests an increase in the length of the micelles or an agglomeration of the rod-like micelles.

The results of the experiment showed that the particle size distributions of NDT-6, NDT-7, and NDT-8 were virtually equal to the particle size distribution of NDT-5. From these results, we believe that the addition of rust inhibitors has little effect on the formation of micelles.

The particle size distributions of NDT-9 and NDT-10 are shown in Fig. 5.37A and B, respectively. One peak can be observed for NDT-9 around 20 nm, and the distributions of particles smaller than 5 nm are small. Conversely, the one peak for NDT-10 occurs around 23 nm. The particle sizes of NDT-9 and NDT-10 are smaller than those of NDT-2 and NDT-5, but slightly larger than those of NDT-3

Fig. 5.38A and B shows the experimental results of the friction factors and the drag reduction ratio, respectively. *Dotted line* in Fig. 5.38A is an MDR asymptote.[45] The viscosity of distilled water, which was used as the solvent, was employed to calculate *Re*. *DR* in Fig. 5.38B is defined as $DR = \{(\lambda_w - \lambda_s)/\lambda_w\} \times 100\%$. λ_w and λ_s are the frictional factors for the distilled water and test fluid, respectively.

It is evident that the Ethoquad aqueous solution (NDT-2) provided drag reduction, as has been previously reported. The drag reduction increased with increasing *Re*, reaching a maximum of approximately 60% at *Re* = 8000. The friction factor for NDT-3, which had particle size distributions significantly smaller than those of NDT-2 approached that for water in the laminar flow region. The drag reduction ceased and the determined friction factor was in agreement with the straight line of the Newtonian turbulent region. Zhang

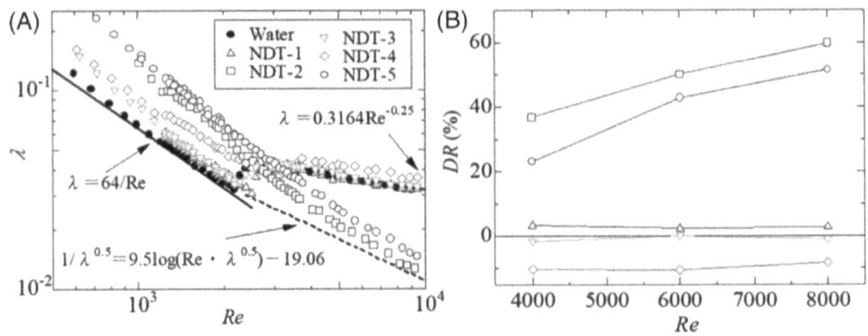

Figure 5.38 ***Drag reduction of surfactant solutions, NDT-1–NDT-5.*** (A) Friction factor and (B) drag reduction ratio.

et al. reported that the smaller micelle size causes a reduction in drag reduction rate. The tendency of these experimental data agrees well with Zhang's data. No drag reduction effect was observed for NDT-1, which was the one in which the particle size distributions could not be determined. On the other hand, drag reduction similar to that of NDT-2 was observed in the turbulent region for NDT-5, which had particle size distributions slightly larger than those of NDT-2. The drag reduction rate was approximately 50% for $Re = 8000$, that is, approximately 10% lower than that for NDT-2.

From the above experimental results, we believe that there is a close relationship between the drag reduction and particle size distribution. In particular, significant reduction of the particle size distribution eliminates the drag reduction effect. Additionally, the results of degradation using the centrifugal pump indicate that the particle size distribution is directly related to the degree of drag reduction. To confirm the relation between the drag reduction and particle size distribution, the pressure loss of NDT-5, NDT-6, NDT-7, NDT-9, and NDT-10 were measured. On the basis of these experimentally determined particle size distributions, it could be predicted that NDT-5, NDT-6, and NDT-7 would have drag reduction capacity, and that NDT-9 and NDT-10 would have little or no drag reduction capacities.

Fig. 5.39A and B shows the effects of adding rust inhibitors such as benzotriazole, amine salts, and tolyltriazole (a metal corrosion inhibitor) to NDT-5 on the friction factor and the drag reduction ratio, respectively. Experimental data show that the maximum drag reduction is obtained in the range as shown in comparison with the hitherto known result. As stated earlier, considering that the particle size distribution of these solutions is the same as that of NDT-5, the drag reduction properties of NDT6, NDT-7, and NDT-8 can be expected to be virtually the same as those of NDT-5. As is clear from Fig. 5.39A, the addition of the rust inhibitors appeared to have little effect

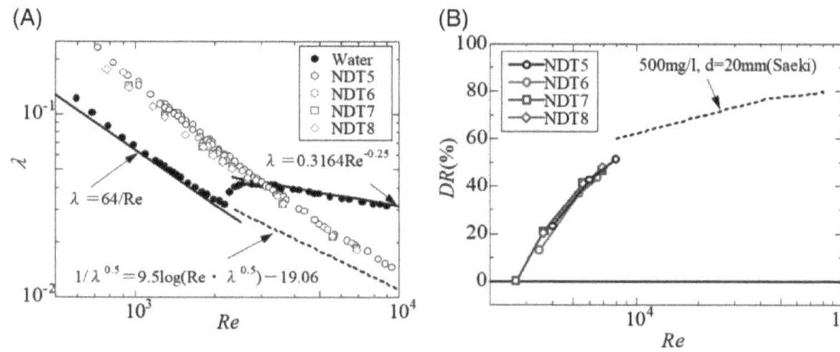

Figure 5.39 ***Drag reduction of surfactant solutions, NDT-5–NDT-8.*** (A) Friction factor and (B) drag reduction ratio.

on the drag reduction. From this result, the relationship between the particle size distributions and the drag reduction effect are also demonstrated.

On the other hand, it is known the surfactant solutions increase the rate of corrosion. It will be necessary to learn more about the effect for actual application.

Table 5.2 presents the results of the corrosion test. The corrosion coupons were made from steel, aluminum, copper, and brass, and each weighed 250 g. The test temperature and time were 80 °C and 168 h, respectively. The weights of each coupon specimen before and after the test are included in Table 5.2. The weight change was an indication of the magnitude of the corrosion of the metal. Corrosion of iron occurred in all the test fluids, whereas those of aluminum, copper, and brass were suppressed by the addition of the rust inhibitors. Owing to the difficulty of performing an experiment on a solution in which the corrosion of a metal is ongoing, the particle size distributions were measured before and after the corrosion. This

Table 5.2 Experimental results on corrosion test

Material	**Weight change before and after the experiment (g)**			
(Size (mm))	**NDT-5**	**NDT-6**	**NDT-7**	**NDT-8**
Steel (25 × 50 × 4)	4.5	9.3	11.2	2.9
Aluminum (25 × 50 × 3)	−7.0	0.1	−0.6	−1.5
Copper (25 × 50 × 1.6)	1.5	−0.03	−0.04	−0.01
Brass (25 × 50 × 1.6)	1.5	−12.5	−0.9	−0.5

could, however, not be done for NDT-6, and its particle size distribution was considered to be unchanged by corrosion. It can therefore be expected that the drag reduction would be maintained by a surfactant aqueous solution containing corroded metal, just as is the case of the fresh solution.

5.3.2 Flow Between Two Coaxial Cylinders

Fig. 5.40A–C illustrates the flow conditions of several test fluids in the gap. The time for the flow to become stable was t_s determined using Snyder's equation[19] as given by

$$t_s \cong 0.15\frac{L^2}{\nu} \tag{5.10}$$

where, L and ν are the length of the cylinder and kinematic viscosity, respectively. In this experiment, the time t_s is about 10,800 s.

As is evident in Fig. 5.40A(iii), six Taylor cells appeared in the case of tap water. Observations of the flow confirmed that the boundaries of the Taylor cells did not fluctuate with time and that secondary flow, which had come into contact with both end planes of the inner cylinder, approached the inner wall from the outer one. And the vortex ring was circumferentially

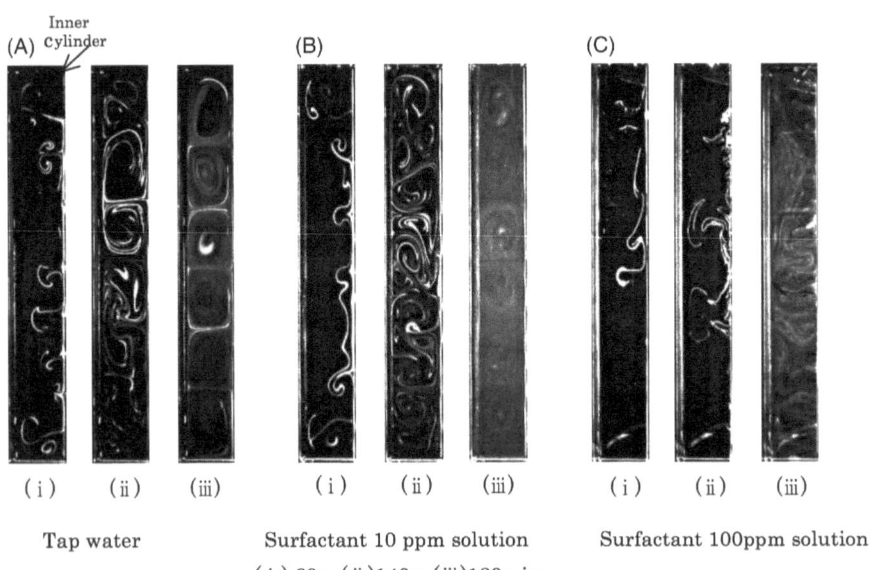

Figure 5.40 ***Formation process of Taylor cells.*** (A) Tap water, (B) surfactant 10 ppm solution, and (c) surfactant 100 ppm solution.

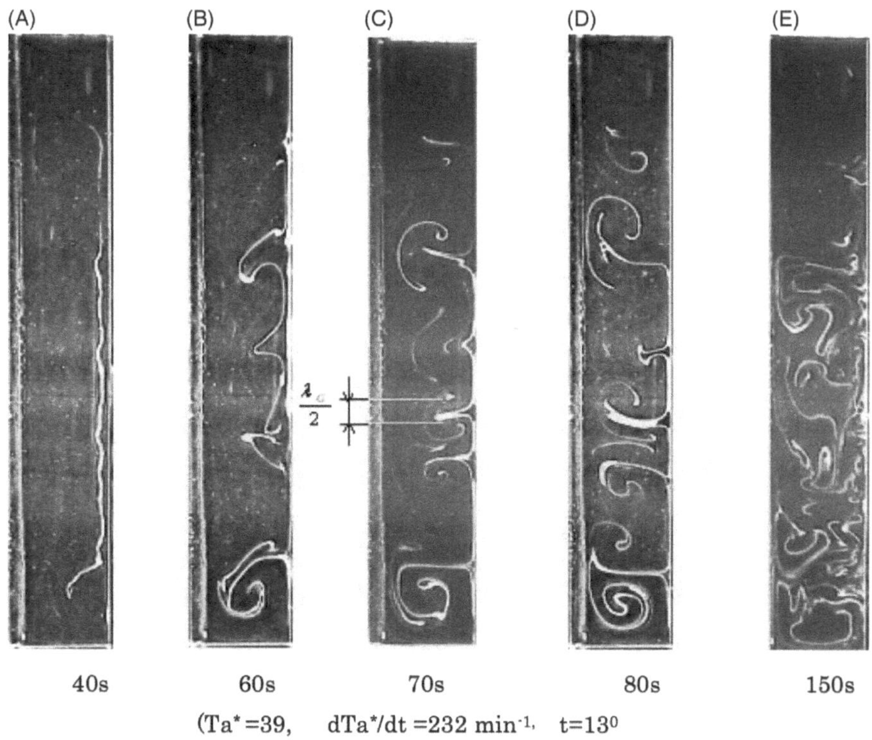

Figure 5.41 ***Formation process of Taylor cells for 100 ppm surfactant solution.*** (A) 40 s, (B) 60 s, (c) 70 s, (D) 80 s, (E) 150 s. (Ta* = 39, dTa*/dt = 232 min^{-1}, t =13°.

uniform. These facts mean that the flow is in a normal mode,[48] which is generally the case for the laminar Taylor vortex flow.

Taylor cells[49] also appeared in the case of the 10 ppm surfactant solution, as shown in Fig. 5.40B, again exhibiting the normal mode. However, no Taylor cells appeared in the cases of 100 ppm solutions shown in Fig. 5.40C even after the stabilization time, t_s.

The detailed formation process can be viewed from Fig. 5.41 after it is confirmed where the phenomenon occurs. The dome (pileus) of the Görtler vortex separates from the main (stipe) part of the vortex in Fig. 5.41C and D. This separation is related to the fact that the Görtler vortex does not grow and the Taylor cell is not able to form as a result, as illustrated in Fig. 5.41E. This separation phenomenon was also observed in 50 ppm surfactant solutions. As stated earlier, a Görtler vortex does not separate in tap water, as shown in Fig. 5.10C and D for Newtonian fluid. The range of the maximum shear rate, calculated by applying the mean velocity of Couette

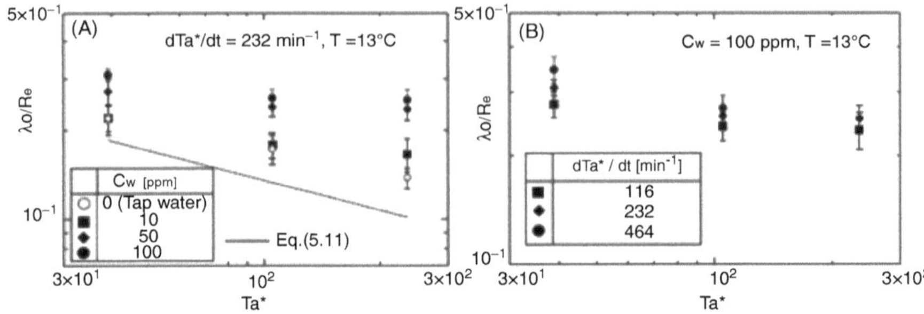

Figure 5.42 ***Görtler wavelength of surfactant solutions.*** (A) Effect of concentration and (B) effect of acceleration.

flow is $3.6 \times 10^2 \leq \dot{\gamma} \leq 8.8 \times 10^2$. As mentioned with respect to the flow curves, these results demonstrate that the higher local viscosity prevents development of the vortex in the surfactant solutions.

Fig. 5.42A and B shows the effect of surfactant concentration on the wavelength of the Görtler vortex. The wavelength was measured by the method ofWei et al.'s, in which the center-to-center distance was measured and two times this value was taken to be the wavelength λ_G. The examples of the definition of λ_G are shown in Fig. 5.41. The experimental values are mean values obtained from 50 to 100 samples, and the error bars in the figure represent 95% confidence levels. The *solid line* in Fig. 5.42A is the experimental result for Newtonian fluids obtained by Wei et al. expressed as

$$\frac{\lambda_G}{R_0} = 31.89\left(\frac{\omega_i^2 R_0^4}{\nu^2}\right)^{-\frac{1}{3}} \tag{5.11}$$

where R_o, ω_i and ν are the outer cylinder radius, the angular velocity of the inner cylinder, and the kinematic viscosity, respectively. Although our experimental data for tap water are above the line, there is a tendency for better agreement at lower Taylor numbers. Wei et al. reported that the wavelength increases with the clearance distance and Eq. (5.11) applies to small gaps. Because of the wide gap in our experiment, the data for tap water lie above the solid line.

The wavelength of surfactant solutions decreases slowly with T_a^* in comparison to tap water for all these concentrations. This means that the solutions' stabilizing effect on the Görtler instability increases with T_a^* and with surfactant concentration. The effect of inner cylinder acceleration on the wavelength is presented in Fig. 5.42B, which shows only a weak dependence. Counter-ion was added to the surfactant solutions, as described in the previous section, affecting the shear-induced structure of the solutions. Experiments are carried out to visualize the formation process of the Taylor cell of the solution without counter-ion because it is known that a cationic surfactant forms long-threadlike micelles in aqueous solution by adding sodium salicylate or cetyltrimethylammonium and counter-ion cause a micelle network to form. The threadlike micelles make concentrated entanglement networks to show pronounced viscoelastic behavior as well as concentrated polymer systems.[50] However, a responsible mechanism for the longest relaxation mode of the threadlike micellar system is different from that of the polymer system. Gupta et al.[51] reported the results of a linear stability analysis of the Taylor Couette of non-Brownian fiber suspensions that irrespective of the closure approximation, used the fiber additives suppress the centrifugal Taylor Couette instability, that is, the critical Reynolds number increases with fiber volume fraction and aspect ratio. The enhanced stabilizing cab is attributed to the fact that the suspension develops negative first and second normal stresses in the flow system. Moreover, the dynamic interaction between turbulence and the viscoelastic behavior play a significant role in the stability.[52] Generally speaking, the influence of the viscoelastic behavior in the shear-induced structure on the stability of the forming process of Taylor cells is relatively less understood at present.

As described earlier in this section, Taylor cells did not appear for 100 ppm solutions with counter-ion added. However, the Taylor cells clearly form in the absence of counter-ion. Thus, this result suggests that increasing the local viscosity due to a shear-induced structure causes a stabilizing effect on the Görtler instability. As a result, Taylor cells do not appear in the 50 and 100 ppm surfactant solution. It is interesting in conjunction with the micelle formation of surfactant solutions and the extinguishing the relatively large vortex in considering the internal structure of the solutions.

5.3.3 Flow Around a Rotating Disk

There have been few studies in the frictional resistance of a rotating disk in surfactant solutions. It is necessary to estimate the minimum moment coefficient in practical applications of a rotating disk in a drag-reducing surfactant solution. Two analytical methods can be used to obtain the minimum values of frictional resistance of a rotating disk. The first method uses the momentum equations; the second involves the application of momentum integral equations for a boundary layer. The first method is not suitable for obtaining the moment coefficient of a rotating disk, as the rheological equation of surfactant solutions has not yet been derived. The second method is applicable, as analysis is possible if the velocity profile of the boundary layer are known, and the logarithmic velocity profiles in pipe flow have been presented in many studies. Additionally, in the case of dilute polymer solutions, the final velocity profile for maximum drag reduction has been obtained. If the velocity profile in pipe flow can be used to obtain the moment coefficient of a rotating disk with maximum drag reduction, the results using the moment integral equation for boundary layer on a rotating disk can be used to estimate the limit of drag reduction due to surfactant solutions. As mentioned earlier, there are little data available pertaining to the velocity profile of a rotating disk flow in a surfactant solution at present. If we can assume the velocity profile in the boundary layer of a rotating disk in surfactant solutions as shown in Fig. 5.43 the moment coefficient are given by Eqs. (5.5) and (5.6).

The limiting maximum drag reduction asymptote for the moment coefficient of a free disk in surfactant solutions based on the experimental results for the moment coefficient of a rotating disk and flow angle be derived as follows.

It is well known that the logarithmic velocity profiles in a circular pipe cover a wide range of Reynolds numbers. If $\tau_0 = \rho v_*^2$ represents shearing

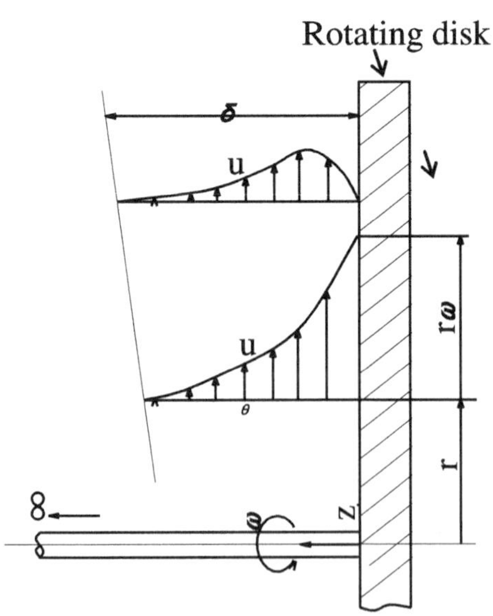

Figure 5.43 ***Flow model of velocity profiles in boundary layer on a rotating disk.***

stress at a wall, u is the velocity near the disk wall, the logarithmic velocity profile is given as

$$u^+ = A\log_e y^+ + B \tag{5.12}$$

where, $u^+ = u/v_*$ and $y^+ = v_* z/\nu$.

The constants A and B are generally known semiexperimentally to be $A = 2.5$ and $B = 5.5$ for Newtonian fluids. Virk et al. obtained the ultimate velocity profiles for dilute polymer solutions, as shown by the following equation, in the maximum drag reduction range for a circular pipe.

$$u^+ = 11.7\log_e y^+ - 17 \tag{5.13}$$

Fig. 5.44 shows the experimental results of velocity profiles for surfactant solution presented in other studies[53–56]. Additionally, considering their experimental data, Zakin et al.[55] proposed the following equation for surfactant solutions:

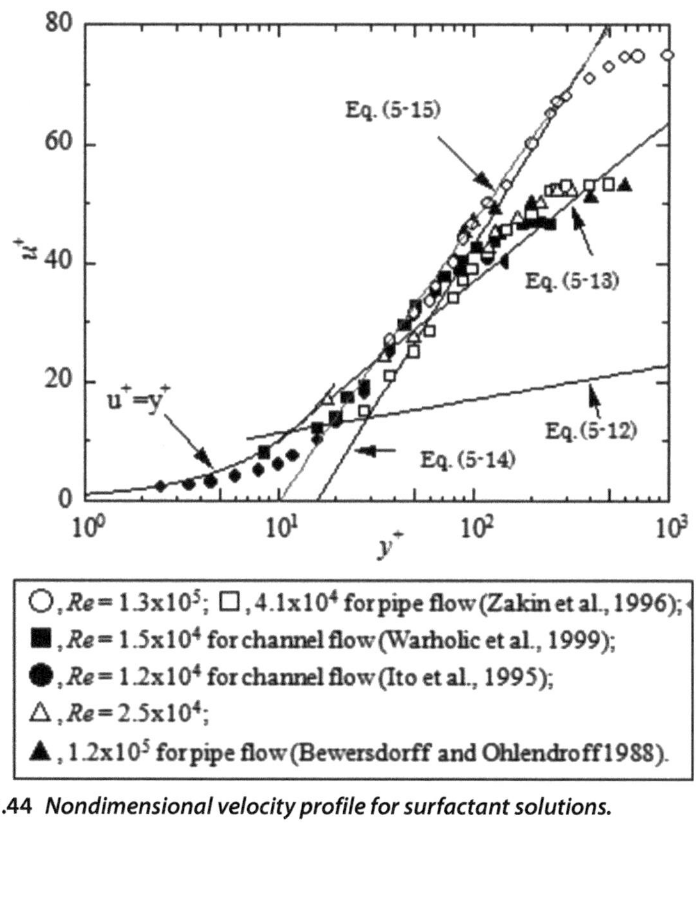

Figure 5.44 ***Nondimensional velocity profile for surfactant solutions.***

$$u^{+} = 23.4 \log_e y^{+} - 65 \tag{5.14}$$

However, as shown in Fig. 5.44, the logarithmic velocity profiles in surfactant solutions are different from Eq. (5.14), and Zakin et al. suggested that Eq. (5.14) was not the ultimate velocity profile. Ogata and Watanabe[57] assumed the following equation for logarithmic velocity profile of surfactant solutions with reference to these experimental data in Fig. 5.44.

$$u^{+} = 20.4 \log_e y^{+} - 47 \tag{5.15}$$

It is clear that Eq. (5.15) agrees well with the experimental data for surfactant solutions when compared with Eq. (5.14) in Figure 5.44.

For Newtonian fluids, based on the logarithmic velocity profile of turbulent pipe flow, Goldstein[23] and Ito[25] reported analytical results for the moment coefficient of a rotating disk by using momentum integral equations of the boundary layer on a rotating disk. With reference to their analysis, the moment coefficient of a rotating disk in drag reducing surfactant solution is calculated by applying Eq. (5.15) for the velocity profiles of the boundary layer on the disk.

If u is the velocity near the rotating disk surface, the radial and tangential components of velocity profiles with no-slip boundary condition are assumed as follows:

$$u_r = \frac{\alpha}{\sqrt{1+\alpha^2}}\left(\omega r\sqrt{1+\alpha^2} + Av_* \log_e \frac{z}{\delta}\right)\left(1 - \frac{z}{\delta}\right) \tag{5.16}$$

$$u_\theta = -\frac{Av_*}{\sqrt{1+\alpha^2}} \log_e \frac{z}{\delta} \tag{5.17}$$

where, δ and α are boundary layer thickness and a constant related to the flow direction, respectively.

The radial and tangential components of shearing stress acting on a disk wall are obtained as

$$\tau_r = \frac{\alpha}{\sqrt{1+\alpha^2}}\tau_0 \; and \; \tau_\theta = -\frac{1}{\sqrt{1+\alpha^2}}\tau_0$$

By substituting Eqs. (5.16) and (5.17) into Eqs. (5.5) and (5.6), we obtain the differential equations for α and δ.

Now, the values of functions α and δ can be assumed as

$$\alpha = \alpha_1 \frac{Av_*}{\omega r} + O\left[\left(\frac{Av_*}{\omega r}\right)^2\right]$$

$$\delta = \delta_0 + O\left(\frac{Av_*}{\omega r}\right)^2$$

where, α_1 and δ_0 are functions of r.

For large Reynolds numbers, the high-order terms involving $(Av_*/\omega r)$ are very small. Thus, these expressions may be reduced by neglecting the high-order terms. The differential equation are solved for α_1 and δ_0

$$\alpha = \sqrt{\frac{3}{2}} \frac{Av_*}{\omega r} \tag{5.18}$$

$$\delta = \frac{4}{15A^2}\sqrt{\frac{2}{3}} r \tag{5.19}$$

The moment acting on one side of the disk is

$$M = -2\pi \int_0^a r^2 \tau_\theta dr = 2\pi \rho a^2 \int_0^\delta u_r u_\theta dz$$

We obtain the moment coefficient acting on one side of the disk from the following simple formula:

$$\frac{1}{\sqrt{C_m}} = m \log_{10}\left(\mathrm{Re}\sqrt{C_m}\right) + n \tag{5.20}$$

where m and n are

$$m = \frac{A}{2}\sqrt{\frac{15\sqrt{3}}{\pi(3\sqrt{3(1+\alpha^2)} - 7\sqrt{2}\alpha)}} \log_e 10$$

$$n = \frac{\sqrt{3}A}{2\sqrt{(1+\alpha^2)}}\sqrt{\frac{5\sqrt{3}(1+\alpha^2)}{\pi(3\sqrt{3(1+\alpha^2)} - 7\sqrt{2}\alpha)}}\left(\log_e\left(\frac{8\alpha}{45A^3}\right) + \frac{B}{A}\right)$$
$$-m \log_e 10 \cdot \log_{10}\left(\frac{2\alpha}{3A}\sqrt{\frac{2\pi(3\sqrt{3(1+\alpha^2)} - 7\sqrt{2}\alpha)}{5\sqrt{3}(1+\alpha^2)}}\right)$$

As mentioned earlier, α is a function related to the flow angle ϕ near the rotating disk, where $\alpha = -\tan\phi$. In this study, the value of α was assumed from experimental results of flow angle on the rotating disk. By substituting parameter α in Eq. (5.20), the moment coefficient of a rotating disk can be calculated exactly.

The values of the moment coefficient of a rotating disk in surfactant solutions for maximum drag reduction for several concentrations of surfactant solutions are plotted in Fig. 5.45. Test surfactant solutions were aqueous solutions of Ethoquad O/12 ($C_{18}H_{35}N(C_2H_4OH)_2CH_3Cl$, Lion Co.) at concentrations of 50, 100, and 200 ppm, and sodium salicylate ($C_7H_5NaO_3$) was added as a counter ion. The experiments were performed with the temperature of surfactant solutions maintained at $t = 18$ and 28 °C.

It can be seen that the data approximates the solid line $C_m = 1.85(s/a)^{-1/2}Re^{-1/2}$ which was obtained by Daily and Nece[58] for the enclosed rotating disk, even if the concentration of surfactant solutions

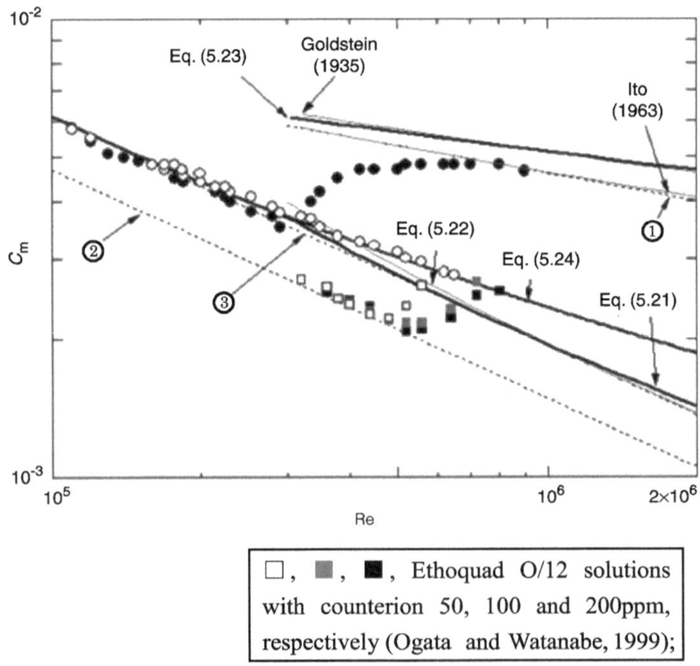

Figure 5.45 ***Moment coefficients of a rotating disk.***

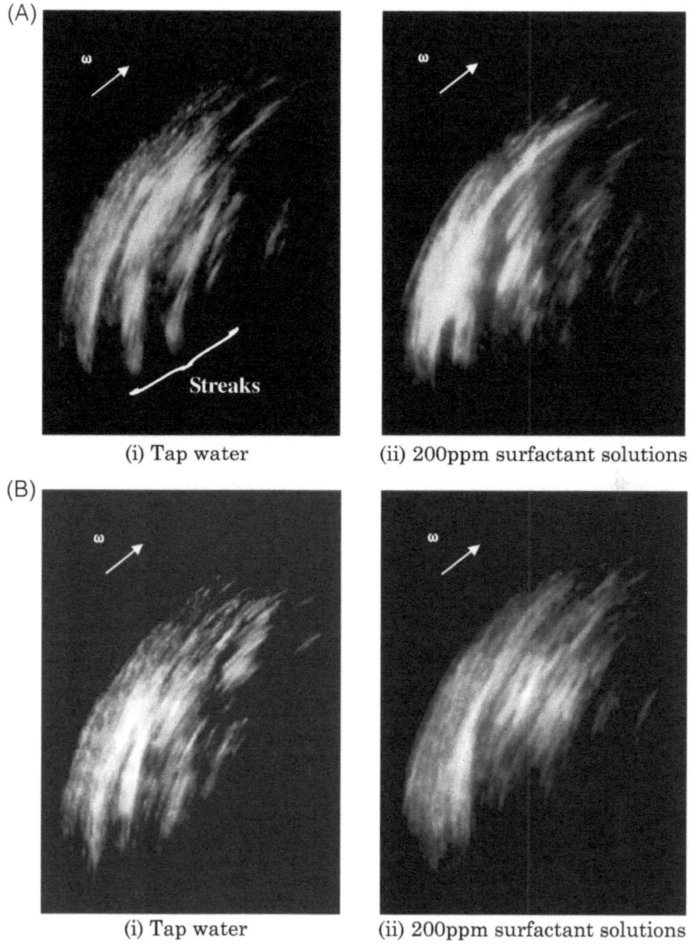

Figure 5.46 ***Flow patterns near rotating disk.*** (A) Re = 2.5×10^5 and (B) Re = 3.5×10^5.

changed. This suggests that the line of the minimum value of the data of surfactant solutions did not vary in this Reynolds number range.

Fig. 5.46A–B shows the flow visualization results from the tracer technique at Reynolds numbers of $Re = 2.5 \times 10^5$ and 3.5×10^5, respectively. The camera shutter speed was 1/60 s. Fig. 5.46A(i) and B(i) shows the flow visualization results for tap water, and Fig. 5.46A(ii) and B(ii) shows the flow visualization results for 200 ppm Ethoquad O/12 solution. The Reynolds number of surfactant solutions was calculated using the viscosity of tap water. In the case of $s/a = 0.115$, Daily and Nece reported also that the transition to turbulent flow began at $Re = 2.8 \times 10^4$ and that turbulence in the boundary layer began at $Re = 1.5 \times 10^5$. Thus, Fig. 5.46A and B represent the transition and turbulent flow ranges, respectively.

For tap water, two or three white streaks in tangential directions, are seen in Fig. 5.46A(i). These streaks represent flow vortices on the rotating disk and appear in the transition to turbulent flow. In the case of Ethoquad O/12 solution shown in Fig. 5.46A(ii), the streaks did not appear clearly. If the contrast between dark and light stripes is strong, the vortices near the rotating disk can be considered to be powerful because the reflected light intensity of the aluminum powder indicates the strength of the flow. Therefore, the vortices in the Ethoquad O/12 solution were weaker than those in tap water. Hence, surfactant solutions reduce the amplitude of the circular vortex on a rotating disk. In addition, the flow direction near the disk is turned toward the outward direction by the surfactant solution.

The streaks are not clear in Fig. 5.46B(i) because the flow was turbulent. In Fig. 5.46B(ii), the contrast between the dark and light stripes of Ethoquad O/12 solution is not stronger than that of tap water. This suggests that surfactant solutions reduce the turbulent flow directional quality.

The flow angle is analyzed by using Eq. (5.8). Fig. 5.47 shows the results for each solution by the solid line. It can be seen that the flow angles of a surfactant solution are larger than those of a Newtonian fluid in the turbulent flow range of the Reynolds number. In other words, the flow direction near the disk is turned outward by the surfactant solution. The calculated results decrease with increasing Reynolds number for surfactant and polymer solutions, but the results for Newtonian fluid remain constant and demonstrated

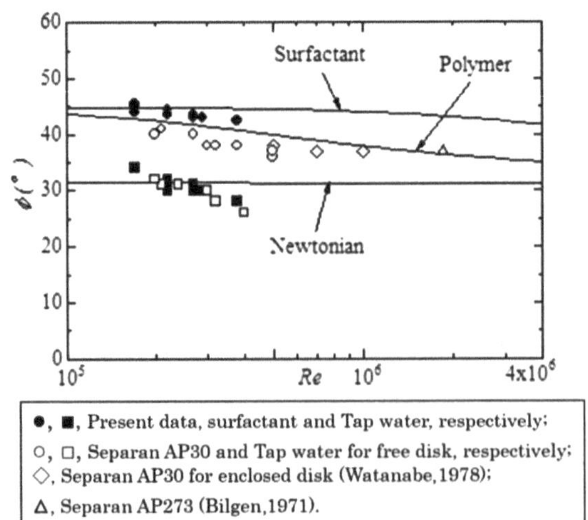

Figure 5.47 ***Flow angle of a rotating disk.***

no dependence on Reynolds number. It is seen that the experimental results for surfactant and polymer solutions agree well with the analytical results in the range $Re < 7 \times 10^5$, and it was clarified that the flow angles for surfactant solution are larger than those for polymer solution. Consequently, increased flow angle on the rotating disk and decreased amplitude of a circular vortex led to drag reduction for the rotating disk due to surfactant solutions.

The limiting maximum drag reduction asymptote for the moment coefficient of a free disk in surfactant solutions based on the experimental results for the moment coefficient of a rotating disk and flow angle will be derived as follows.

As shown in Fig. 5.47, the flow angle on the disk for surfactant solutions decreases gradually with increasing Reynolds number. However, for simplification, we assume a flow angle of $\phi = 45°$ from the experimental and analytical results such that $\alpha = -1$.

By substituting the coefficients $A = 20.4$; $B = -47$ of Eq. (5.15) and $\alpha = -1$ into Eq. (5.20), we obtain the values $m = 16.2$ and $n = -74.7$. Goldstein suggested that it is necessary to modify the values of m and n by experimental data, as the velocity profiles of the boundary layer are approximated by the velocity profiles of circular pipe flow.

Little experimental data exists for a rotating disk in the case of surfactant solutions in large tank. It is well known that the value of the moment coefficient of an enclosed rotating disk decreases by approximately one-half as the fluid in the chamber rotates at approximately one-half of the angular velocity of the disk in this Reynolds number range. Therefore, the experimental data for a rotating disk in a large tank are corrected with those for an enclosed rotating disk.

When the coefficient is corrected considering the experimental result of surfactant solution approximates the *solid line* in Fig. 5.45, we obtain the formula of the limiting maximum drag reduction asymptote for the moment coefficient of a rotating disk in surfactant solutions.

$$\frac{1}{\sqrt{C_m}} = 16.2\log_{10}\left(\mathrm{Re}\sqrt{C_m}\right) - 52.6 \tag{5.21}$$

Eq. (5.21) is shown in Fig. 5.45. It is seen that Eq. (6.21) agrees with data for surfactant solutions, even if the parameter m is not modified. As mentioned earlier, Eq. (6.21) is not directly compared with the experimental results of the enclosed rotating disk. In Fig. 5.45, it seems that Eq. (5.21) shows the behavior of the moment coefficient of the free disk in surfactant solution.

Table 5.3 Comparison of parameters in Eq. (5.20). $\frac{1}{\sqrt{C_m}} = m\log_{10}\left(\mathrm{Re}\sqrt{C_m}\right) + n$

	Surfactant		Polymer	Newtonian		
	Eq. (5.21)	**Eq. (5.22)**	**Eq. (5.24)**	**Eq. (5.23)**	**Goldstein[23]**	**Ito[25]**
m	16.2	18.7	10	2.4	2.78	3.51
n	−52.6	−64.0	−26.2	2.3	0.46	−2.26

For comparison, in the case of the velocity profile of surfactant solutions given in Eq. (5.14), the same calculation procedure was followed. These calculated results are shown in Table 5.3, and the formula using Eq. (5.14) was as follows:

$$\frac{1}{\sqrt{C_m}} = 18.7\log_{10}\left(\mathrm{Re}\sqrt{C_m}\right) - 64 \tag{5.22}$$

It is seen in Fig. 5.32 that the moment coefficient of Eq. (5.22) is smaller than that of Eq. (5.21) in the large Reynolds number range, but the difference may be very small in the Reynolds number range $Re < 2 \times 10^6$.

On the other hand, the moment coefficient of a rotating disk in Newtonian fluids was calculated in the same manner as that in surfactant solution. By substituting the coefficients of Eq. (5.12) into Eq. (5.20), and assuming a flow angle of $\phi = 30°$, we obtain $m = 2.4$ and $n = -2.0$. Modifying the coefficient of n using Kármán's turbulent flow formula[22] $C_m = 0.073Re^{-1/5}$, we obtain the moment coefficient for Newtonian fluid as follows:

$$\frac{1}{\sqrt{C_m}} = 2.4\log_{10}\left(\mathrm{Re}\sqrt{C_m}\right) + 2.3 \tag{5.23}$$

Eq. (5.23) is shown in Fig. 5.45. Goldstein and Ito obtained the formulas of the moment coefficient for a Newtonian fluid by modifying both coefficients m and n based on experimental data. Table 5.3 shows the comparison of these coefficients. The formulas are also shown in Fig. 5.45. It is seen that Eq. (5.23) agrees well with Goldstein's formula in the range of $10^5 < Re < 2 \times 10^6$.

To ascertain the significance of the maximum drag reduction asymptote for surfactant solutions, it is necessary to obtain a similar equation for a dilute polymer solution in which drag reduction of a rotating disk occurred by way of Eq. (5.21). For dilute polymer solutions, Watanabe reported experimental data of the moment coefficient of a disk in a large tank. The experimental results of the maximum drag reduction are shown in Fig. 5.45. The flow angle of polymer solution is assumed to be $\phi = 40°$ from experimental results in previous studies and results presented in this study. By substituting Eq. (5.13) and $\phi = 40°$ into Eq. (5.20), we obtain $m = 10$ and $n = -37$. Modifying the coefficient n using experimental results for maximum drag reduction, we also obtain the formula of the limiting maximum drag reduction asymptote for the moment coefficient of a rotating disk in a dilute polymer solution.

$$\frac{1}{\sqrt{C_m}} = 10\log_{10}\left(\mathrm{Re}\sqrt{C_m}\right) - 26.2 \tag{5.24}$$

As shown in Fig. 5.45, good agreement between experimental data and Eq. (5.24) is obtained in the range $Re < 7 \times 10^5$.

It can be seen that the moment coefficient for surfactant solutions decreases compared with that of dilute polymer solutions, and the tendency is similar to that of pipe flow. As we can decide the numerical value of m in Eq. (5.20) by the decision of the flow angle, the value of m does not require modification by experimental results.

This suggests that the slope of the moment coefficient can be calculated directly using the sloop of the logarithmic velocity profiles of pipe flow. Thus, it can be considered that the slope of the moment coefficient represents the limit of the drag reduction due to surfactant solutions.

5.3.4 Flow Past a Circular Cylinder

Generally, at low shear rates the surfactant solutions are slightly shear thinning and the micelles become more aligned in the flow direction; as a result the shear viscosity increases suddenly and the solution has a viscoelasticity. This larger superordered structure of the aligned micelles is referred to as the shear induced state (SIS). Drag reducing surfactant solutions generally have strong viscoelasticity, so that viscoelasticity is considered to play an important role in the drag reduction effect. However, Lin et al.[59] investigated the pipe flow of surfactant solutions (Arquad 16-50: cetyltrimethyl ammonium chloride, Ethoquad O/12: oleyl-methyldihydroxyethyl ammonium chloride) with excess addition of counter ion (NaSal), and reported that the turbulent drag reduction is also shown by surfactant solutions that do not show viscoelasticity. Considering the drag reducing surfactant solutions not always have viscoelasticity as mentioned earlier, there are many unclear points for the relationship between the viscoelasticity and the drag reduction of surfactant solution at present.

On the other hand, by studying the flow past a circular cylinder, the effects of flow separation and wake on the drag of cylinder can be examined. In addition, such flow is related to the shear flow, which produces the shear induced state, as mentioned earlier. Therefore, to clarify the effect of the shear induced state on the flow characteristics and to examine the mechanism of drag reduction, the flow of a cylinder in drag reducing solutions should be investigated.

For drag reducing polymer solutions, James and Acosta[60] measured the drag of cylinder in a dilute polymer solution (Polyethylene oxide, WSR-205, WSR-301) in the $Re < 50$ region and clarified that the drag coefficient is independent of the uniform velocity and becomes constant above the critical uniform velocity, and that the drag coefficient is one order higher than that of water. Koniuta et al.[61] pointed out these phenomena are caused by the existence of a wide stagnation zone around the cylinder. Marero and Mena[62] also showed that this zone becomes 10 times wider than that for water and occurred when the Deborah number, De, reaches 1.

Usui et al.[63] reported that the Strouhal number decreases at $50 < Re < 10^2$ by the use of a polymer solution (Polyethylene oxide, PEO). In contrast, no difference was found[64,65] between the Strouhal number of polymer solutions and that of water at $10^2 < Re < 10^4$. In the case of the Reynolds numbers above 10^4, Turgut et al.[64] conducted an experiment involving a dilute polymer solution (Polyethylene oxide, WSR-301) at $5 \times 10^4 < Re < 3 \times 10^5$ and found a drag reduction for the circular cylinder. They showed that the decrease in drag was due to the reduction in the size of the wake. The strong destabilizing effect of the additives on the vortex breakdown is also described. A similar study by Luikov et al.[66] has shown that the drag reduction is partly due to a change in the characteristics of the vortex motion behind the cylinder.

With respect to the surfactant solutions, however, few studies have examined the flow around a circular cylinder. Research on surfactant solutions in an external flow is difficult due to bubble formation and complicated rheological behavior. Bergin et al.[67] investigated the flow past a circular cylinder in dilute surfactant solutions (Tetradecyl trimethyl ammonium bromide with sodium salicylate, TTAB +NaSal.) at $10^2 < Re < 2 \times 10^3$ by LDV measurements and by visualization. They showed that the frequency of the vortex shedding of surfactant solutions decreases at $2 \times 10^2 < Re < 2 \times 10^3$,

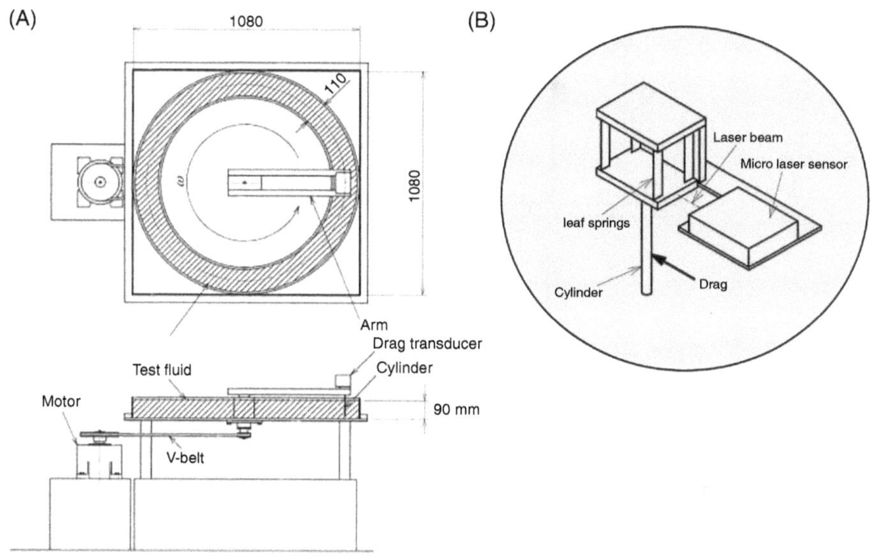

Figure 5.48 ***Experimental apparatus for 10 < Re < 2 × 10²***. (A) and (B) detail of drag measurement device.

and that the flow pattern became smooth when a shear-induced state was formed. They also showed that the stream-wise components of the turbulent velocity fluctuations in dilute surfactant solutions were greater than those in tap water, whereas the tangential components were reduced dramatically. However, they did not measure the drag of the cylinder, and the relationship between the flow pattern and the drag is not known at present. In this section, it will be cleared the effect of the drag reducing surfactant additives on the flow characteristics around the cylinder at $10 < Re < 10^5$ by measuring the drag and by visualizing the flow experimentally.

Figure 5.48A and B [68] shows the experimental apparatus used for drag measurement at $10 < Re < 2 \times 10^2$. Fig. 5.48A and B shows a schematic diagram of the apparatus and a detailed illustration of the drag detection unit, which can measure directly the drag acting on the circular cylinder, respectively. The test cylinder, having a diameter of 2 mm, was fixed on a flat plate in a vertical position. The flat plate was suspended horizontally by two leaf springs. The leaf spring is made of phosphor bronze 0.05 mm in thickness, 74 mm in length, and 15 mm in width. The flat plate was slightly displaced in a direction perpendicular to the axis of the cylinder when the drag force acted on the cylinder. This small displacement was detected by a laser displacement sensor and was converted into the drag using the calibration factor of the leaf springs. The drag detection unit was

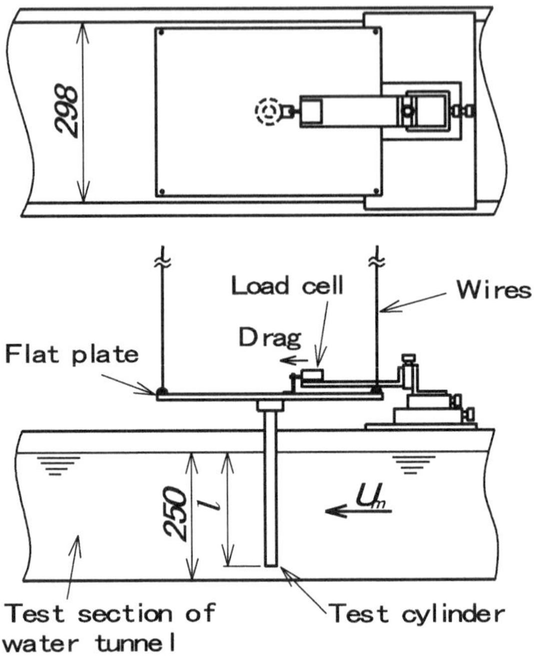

Figure 5.49 ***Experimental apparatus for $2 \times 10^2 \leq Re \leq 7\times \leq 10^3$.***

calibrated by applying a known static load of the weight to the cylinder, and nonlinearly error was 0.35%. The circular towing tank was made of stainless steel of 860 mm in inner diameter, 1080 mm in outer diameter, and 90 mm in depth. The drag detection unit and the test cylinder were mounted on the rotating arm, and moved in the circular towing tank filled with the test fluids. The rotational speed was adjusted by controlling the motor unit.

For the $2 \times 10^2 < Re < 10^5$ region, the experiments were performed in a vertical recirculating water tunnel equipped with a clear plastic test section 298 mm in width, 250 mm in height, and 1190 mm in length as shown in Fig. 5.49. In the water tunnel, the combination of a motor with continuously variable speed and a pump circulated the fluid at test section velocities ranging of up to 1.2 m/s. It is confirmed that the uniform velocity is obtained within ±1.51% for the mean velocity. The cylinders with d = 20, 35, and 50 mm, extending through both tunnel walls, were used for the pressure measurement and the measurement of the velocity of the wake at $10^4 < Re < 10^5$. The pressure distribution around the cylinder was measured by means of a pressure transducer. A static pressure hole of diameter 0.5 mm was located at a distance 400 mm upstream of the cylinder. A

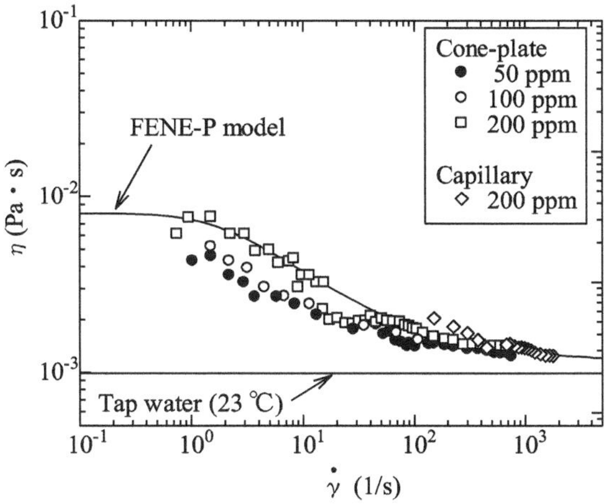

Figure 5.50 ***Viscosity of surfactant solutions.***

pressure hole of diameter 0.5 mm was bored through the cylinder surface at a point midway along the length of the cylinder and was connected to a pressure transducer from the upper side of the cylinder. The local pressure on the cylinder surface was measured at 5-degree intervals. The flow patterns are visualized using a tracer of water-based paint.

The test fluids were tap water and aqueous solutions of oleyl-bihydroxyethyl methyl ammonium chloride ($C_{18}H_{35}N(C_2H_4OH)_2CH_3Cl$, trade name: Ethoquad O/12) at concentrations of 50, 100, and 200 ppm. Sodium salicylate (NaSal) was added as a counter-ion. The concentration ratio of counter-ion to surfactant was set at a molar ratio of 1:1. The viscosity of the surfactant solutions was measured using capillary viscometer, and the data are presented in Fig. 5.50. The curve of the FENE-P model used in the simulation is also shown in the figure. The shear viscosity increases with solution concentration and decreases with shear rate. However, the shear viscosity tends to be constant in the high-shear-rate region. The viscosity differed with the type of viscometer,[69] and the hysteresis of the viscosity occurred when using a cone-plate viscometer.[70] As the viscosity of surfactant solutions is dependent on several parameters, the Reynolds number was calculated based on the viscosity of tap water.

By making the approximation that the total drag almost equals the pressure drag, the drag coefficient C_d of cylinder was obtained by integrating the pressure coefficient C_p as following equation:

$$C_d = \frac{2D}{d\rho U_m^2} = \int_0^{\pi} C_p \cos\phi \, d\phi \tag{5.25}$$

To measure the velocity behind the cylinder, a hot-film probe was positioned downstream at half-length of the cylinder. Measurements were made at seven downstream positions; $x = 30$, 60, 90, 120, 150, 180, and 210 mm, and at 25 traverse stations behind the test cylinder. James and Acosta reported that calibration of the hot-film probe is necessary when used in dilute polymer solution because of the influence of polymer additives on the probe. The calibration appears to be also necessary for the probe when used in the surfactant solution. Therefore, the correlation between the experimental data for tap water and that for surfactant solutions is thought to require some correction. The calibration of the hot-film probe was performed directly using the steady rotating channel flow of surfactant solutions at the same concentration and temperature. In addition, the drag coefficient was calculated based on the velocity defect for comparison with that obtained based on the pressure distribution Eq. (5.25). The total drag of the cylinder in two-dimensional flow can be calculated by the weighing lacks of velocity profiles. The total drag per unit length on a circular cylinder in an uniform velocity U_m at the front of a cylinder is given from the momentum method as follows:

$$D = \rho \int_{-\infty}^{+\infty} U(U_m - U)\,dy \tag{5.26}$$

where, U is the velocity at cylinder wake.

In addition, we visualized the flow around the cylinders to clarify the relationship between the drag and the flow pattern past a circular cylinder. The test cylinders had $d = 2$ and 20 mm. For the cylinder with $d = 2$ mm, the visualization was performed using the laser-induced fluorescence technique. In place of the drag detection unit shown in Fig. 5.35, a digital video camera was placed on the arm. For the cylinder with $d = 20$ mm, the flow was visualized via a water-based dye injected upstream of the cylinder, which was placed in the test section of the recirculating water tunnel.

Fig. 5.51 shows the experimental results for the drag coefficients in surfactant solutions and tap water, respectively. For comparison, the experimental results for a Newtonian fluid with $(l/d) = 5$ and $(l/d) = \infty$ obtained by Wieselsberger[71] are also shown in this figure. The drag coefficient from the velocity profiles and that from the pressure distribution agree within approximately 3.5% for both tap water and surfactant solutions. The results for the drag coefficient in tap water agree well with Wieselsberger's data. For tap water, therefore, it can be considered that there are few effects of the flow geometry and the experimental methods on the drag coefficient.

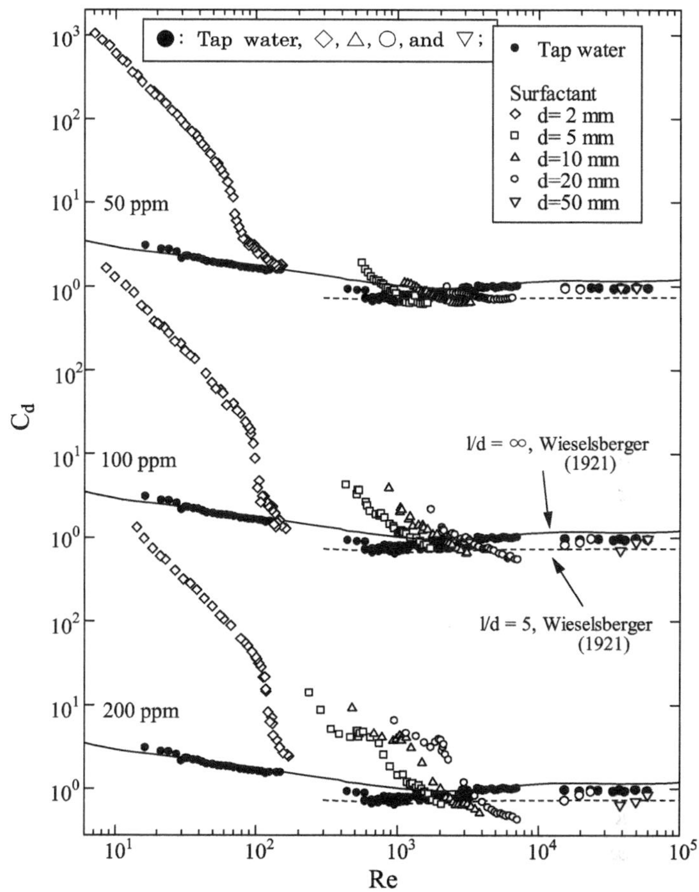

Figure 5.51 ***Drag coefficients of surfactant solutions.***

On the other hand, the drag coefficients of surfactant solutions are significantly different from that of water, and the data is significantly dependent on the cylinder diameter in Fig. 5.51.

For the cylinder with d = 20 mm, the drag coefficient was higher than that of tap water at $10^3 < \mathrm{Re} < 3 \times 10^3$ for 200 ppm surfactant solution. The drag coefficient decreased with increasing Reynolds number, and the slope of the curve of the drag coefficient increased with the concentration. When the Reynolds number reached approximately 3×10^3, the drag coefficient decreased compared to that of tap water. In other words, drag reduction occurred in this Reynolds number range. The maximum drag reduction rate reached up to 55% at Re = 7×10^3 in 200 ppm surfactant solution. The onset Reynolds number of drag reduction was almost the same, regardless of the concentration. The drag coefficient decreased with increasing Reynolds number. When the Reynolds number increased further, the difference between the drag coefficient of the surfactant solution and that of water decreased and the drag reduction eventually vanished. These Reynolds number ranges are dependent on the cylinder diameter, and decrease with decreasing the cylinder diameter.

In the case of the cylinder with d = 2 mm, the drag coefficient was 1–3 orders of magnitude higher than that of water at $10 < Re < 10^2$. This higher drag coefficient in the low Reynolds number range gradually decreases with increasing Reynolds number. The drag coefficient also increased with the increase in surfactant concentration in this Reynolds number range. When the Reynolds number reaches approximately 10^2, the drag coefficient has an inflection point, after which the drag coefficient suddenly decreases and approaches the results for tap water at $Re > 10^2$. These inflection points shifted to the high Reynolds number range with increasing cylinder diameter in 200 ppm surfactant solution.

Fig. 5.52 shows the effect of (l/d) on the drag coefficients at $Re = 10^3$ and 2×10^3, respectively. For comparison, the experimental results for a Newtonian fluid with $(l/d) = 5$ and ∞ at $Re = 10^3$ are also shown in this figure. In Fig. 5.52, the drag coefficient of tap water decreases with decreasing (l/d), and the tendency of the data is the same as that of the previous experimental result for Newtonian fluid. In the case of surfactant solution, the drag coefficient decreased with increasing (l/d), different from the case of Newtonian fluid. Because the flow condition which passed behind the cylinder through the cylinder end in surfactant solution could be different from that in tap water. However, the details of the flow condition are

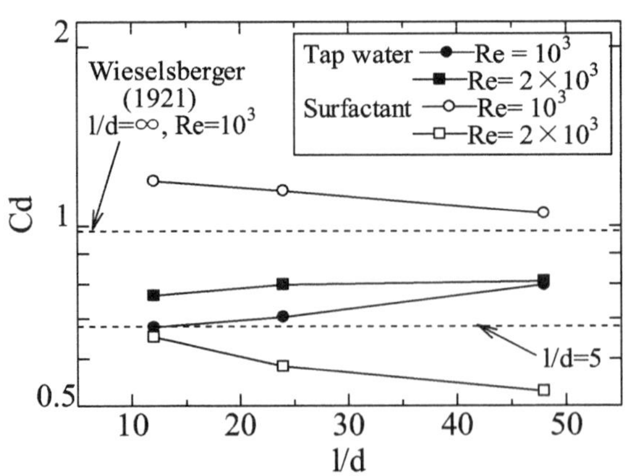

Figure 5.52 ***Effect of (l/d) on drag coefficient for*** **d** ***= 5 mm.***

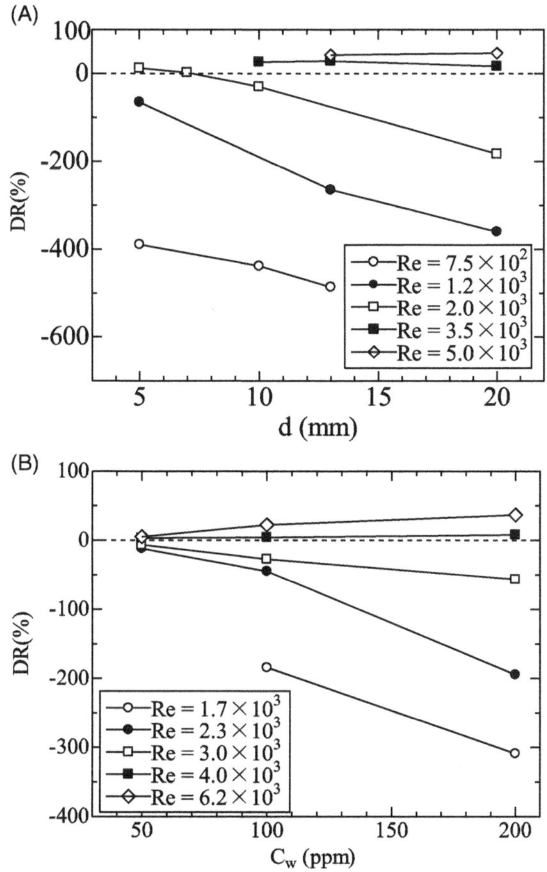

Figure 5.53 ***Drag reduction ratio of a circular cylinder.*** **(A) Effect of a circular cylinder diameter and (B) effect of concentration (*d* = 20 mm).**

not known at present. It is a future problem to clarify the details of the condition.

Fig. 5.53A and B shows the effects of the cylinder diameter and the surfactant concentration on the drag reduction ratio DR, which is defined by $DR = \left\{(C_d|_N - C_d|_S)/C_d|_N\right\} \times 100(\%)$. The subscripts S and N denote surfactant solution and tap water, respectively. It was clarified that DR decreases with increasing cylinder diameter for a small Reynolds number, and that the dependence on the diameter increases with decreasing Reynolds number. For a large Reynolds number, DR increases with increasing diameter. The drag reduction ratio decreases with increasing surfactant concentration in the case of either a small diameter or a small Reynolds number. The effect of the concentration on DR decreases with increasing Reynolds number.

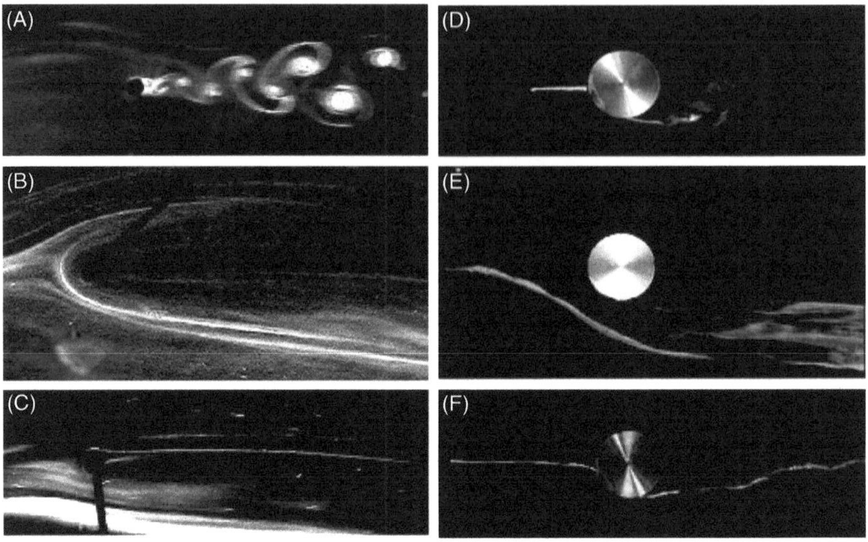

Figure 5.54 ***Flow around a circular cylinder.*** (A) Tap water: Re = 100 (d = 2 mm), (B) 200 ppm surfactant solutions: Re = 20, (C) 200 ppm surfactant solutions: Re = 100, (D) tap water: Re = 10^4(d = 20 mm), (E) 200 ppm surfactant solutions: Re = 2, and (F) 200 ppm surfactant solutions: Re=10^4.

On the other hand, for a large Reynolds number DR increases with concentration. In other words, the increase in the concentration of the surfactant solution amplifies the characteristics of drag reduction or drag increase.

Fig. 5.54 shows photographs of the flow around a cylinder of d = 2 and 20 mm in a 200 ppm surfactant solution, respectively. Generally, for Newtonian fluids, a twin vortex is formed at 10 < Re < 50 and the Kármán vortex street is generated in Re > 50. In Fig. 5.54A and D, the Kármán vortex is clearly observable in the case of tap water in this Reynolds number range.

For surfactant solutions, the flow pattern was significantly different from a Newtonian flow. In the case of the cylinder with d = 2 mm at Re = 20, at which the drag of the surfactant solution increased greatly, the low-velocity zone spreads widely upstream and over the cylinder surface (Fig. 5.54B). This zone becomes 2.5 times of diameter in front and side of the cylinder, and 150 times of diameter stream wise direction, respectively. As the Reynolds number increased, this zone became narrower, until finally becoming nonexistent, and a smooth-flow pattern was observed (Fig. 5.54C). In other words, the drag coefficient decreases with decreasing the size of the low velocity zone.

When the cylinder diameter is increased, a similar flow pattern is observed. For the cylinder with $d = 20$ mm, a wide stagnation zone exists around the cylinder at $Re = 2 \times 10^3$ (Fig. 5.54E), at which the drag of the surfactant solution increases. As in the case of the cylinder with $d = 2$ mm, this zone finally became nonexistent and a smooth-flow pattern was observed (Fig. 5.54F), and the Kármán vortex street (cf. Fig. 5.54D) was not found. For $d = 20$mm, the width of the wake decreased compared with that of tap water for the Reynolds number at which drag reduction occurred. In addition, the point at which the flow separated can be also discerned from Fig. 5.54D and F. Fig. 5.54D shows that the angle of separation for tap water is approximately $\phi = 80^{\circ}$, which is consistent with the experimental data from previous studies with Newtonian fluids. Fig. 5.54F shows that for surfactant solutions the separation point moves downstream. The separation angle is approximately $\phi = 110^{\circ}$ in this case. An increase in the separation angle, due to long-chain molecules, was also found by Brennen[72] and Turgut et al.[64].

To examine the factor of the spread of the stagnant area around the cylinder, the flow around the circular cylinder with $d = 2$ mm was simulated numerically. Here, we used the POLYFLOW (Fluent Inc.) finite-element-based simulation program and the FENE-P model[68]. The FENE-P model is a viscoelastic model, but the model cannot model the SIS. The calculation fluid is a viscoelastic fluid equivalent to 200 ppm Ethoquad O/12 solutions with respect to physical properties of increased apparent shear viscosity.

Fig. 5.55A and B shows the results of the numerical simulation for the Newtonian model and the FENE-P model around the cylinder, respectively. To show that the extent of the streamline of the FENE-P model is very wider than the Newton model, the scales of the two figures are different. For Newtonian model, a twin vortex is formed in the wake (Fig. 5.55A). For FENE-P model,[73] the low-velocity zone spreads widely upstream and over the surface of the cylinder (Fig. 5.55B). The flow pattern of upstream of the cylinder agrees qualitatively with the results observed in flow visualization (Fig. 5.55B), though it does not agree in the wake region. The numerical simulation result seems to indicate that the viscoelasticity, which related to the increased apparent shear viscosity, is one of the factors of the expansion of stagnation zone of upstream of the cylinder.

The results of the velocity profiles of the wake and the pressure distribution around the cylinder at the Reynolds number, in which drag reduction occurs, are shown in Figs. 5.56 and 5.57, respectively. Fig. 5.56 shows the experimental results of the traverse velocity profile at downstream stations

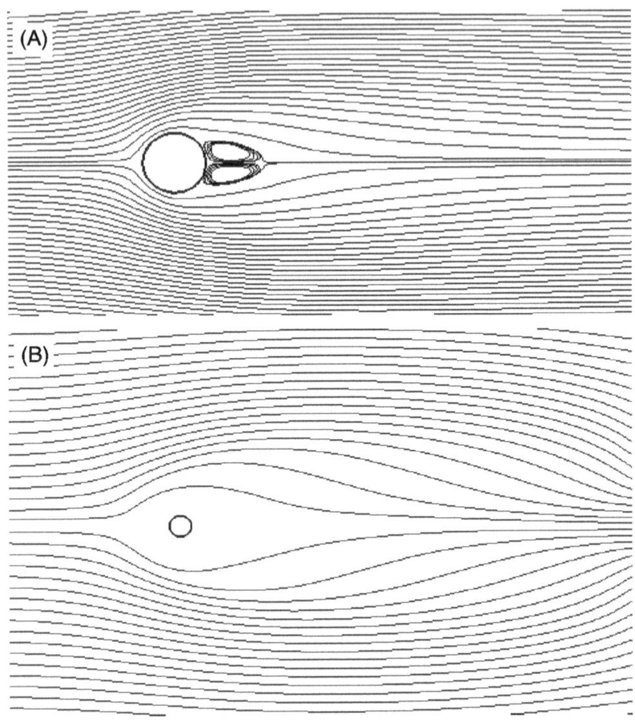

Figure 5.55 ***Numerical simulation results for flow around a circular cylinder.*** (A) Newtonian fluid model; Re = 20 and (B) FENE-P model; Re = 20 (d = 2 mm).

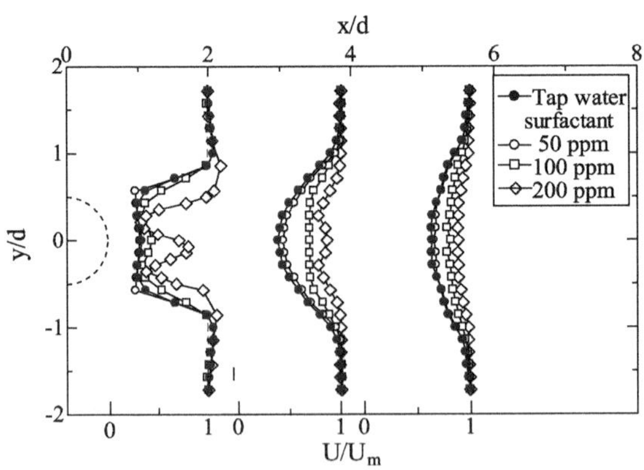

Figure 5.56 ***Velocity profiles in the wake of a cylinder with*** **d** ***= 35 mm at Re = 10^4.***

of x/d = 1.8, 3.5 and 5.1 at Re = 10^4. As seen in the figure, there is only a small difference between the wakes of the 50 ppm surfactant solution and tap water. However, for the 100 and 200 ppm surfactant solutions, the width of the velocity defect is smaller than that for tap water. The velocity profiles

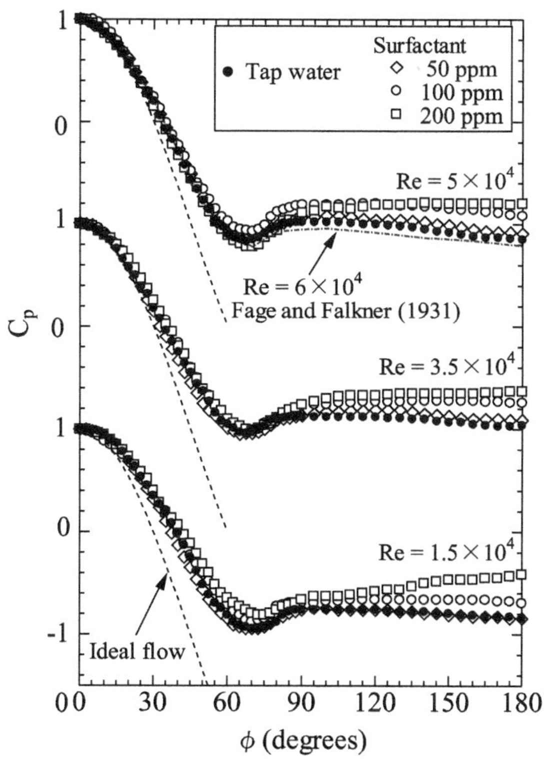

Figure 5.57 ***Pressure profiles of a circular cylinder.***

for surfactant solutions recovered to the mean velocity profiles earlier than did that for tap water. It is interesting that there is a local peak in the center of the velocity profile only for the 200 ppm data.

Fig. 5.57 presents the pressure coefficients for the four test fluids at several Reynolds numbers. The *dashed lines* show the pressure coefficient under ideal flow conditions. The experimental results for $Re = 6 \times 10^4$, given by Fage and Falkner,[74] are also indicated by a *dotted line* for comparison. The tap water data at $Re = 5 \times 10^4$ agree approximately with the experimental results of Fage and Falkner. There is a difference between the surfactant solution and tap water at higher Reynolds numbers, at which the drag reduction occurs, and the difference increases with Reynolds number. The pressure coefficients of the surfactant solution increase with concentration, especially toward the rear of the cylinder.

At lower Reynolds numbers, the drag of the cylinder increases compared to that of tap water, because the wide stagnation area around the cylinder causes the cylinder diameter to increase virtually. The threadlike micelles in the surfactant solution are entangled and form the large network structure (SIS) under shear flow. When the velocity of the flow around the cylinder is very low (or when Reynolds number is very small), the threadlike micelles seem to be entangled and accumulated in front of the cylinder, forming a SIS in the shear region near the cylinder. As indicated by the numerical simulation of Fig. 5.55B, it can be considered that these phenomena depend on the viscoelasticity, which related to the generation of the SIS. With respect to the SIS, Lu et al.[75] confirmed that the micelles network structures exist even though shear is only applied for a very short time from their Cryo-TEM images of Ethoquad O/12 solutions. In this study, the micelles network structures seem to be formed in the shear region of the cylinder.

At higher Reynolds numbers, the width of the wake of the surfactant solution is very narrow, so that the pressure difference between the front and rear of the cylinder decreases as mentioned in Figs. 5.56 and 5.57. In other words, the drag reduction occurs due to the surfactant solution at higher Reynolds numbers ranges, at which the pressure drag makes up the greatest portion of the total drag. When the velocity increases and the force, which works on the SIS increases as a result, the sedimentary layer around the cylinder seems to become thin, because the local viscosity decreases with increasing shear rate. In addition, the effect of extensional flow in the wake of the cylinder seems to be intensified because of the increased elongational viscosity of the SIS.[76] The micelles network structure seems to be extended in the flow direction, so that the cylinder wake can be considered to become narrower and the separation point moves rearward.

The results for the drag coefficient and the flow visualization show that the flow of the surfactant solution is dependent on the cylinder diameter and the mean velocity and that the flow pattern is closely related to the drag characteristics of the cylinder, because the time in which the fluid has been accumulated in neighborhood of the cylinder and the deformation velocity of the fluid are dependent on flow velocity and diameter. In addition, the parameter U_m/d is related to the elongational viscosity of the solutions. Therefore, to clarify the effect of deformation rate (elongational strain rate) on the drag of the cylinder, the experimental results are rearranged using U_m/d.

Fig. 5.58 shows the relationship between U_m/d and the drag reduction ratio. For the cylinder with $d = 20$ mm, drag reduction occurs when $8 < U_m/d < 60$. DR decreases with increasing U_m/d in the range

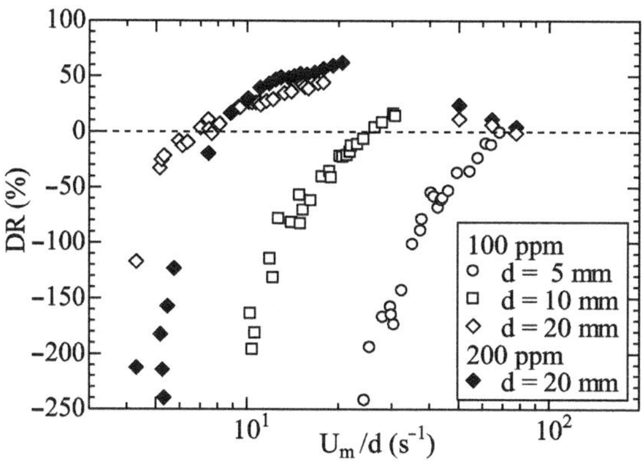

Figure 5.58 ***Effect of Um/d on drag reduction ratio.***

of $U_m/d < 20$, and DR decreases with decreasing U_m/d in the range of $U_m/d > 50$. The onset and end Reynolds numbers of drag reduction are approximately the same, regardless of the concentration. Fig. 5.58 also shows that the U_m/d at which drag reduction begins decreases with increasing cylinder diameter, and that the slope of the DR curve is independent of the diameter. This figure indicates that the drag of the cylinder in the surfactant solution cannot be discussed based only on U_m/d.

To clarify the characteristics of the drag coefficient and flow pattern in surfactant solution in greater detail, it is necessary to accumulate not only more data but also to establish the constitutive equation which can express viscoelastic property of surfactant solution systematically. The experimental drag data and flow visualization results will be useful for the establishment of the constitutive equation of surfactant solutions. Of particular interest this result is the fact that the drag coefficient of surfactant solutions increases compared to that of tap water in the lower Reynolds number range. On the contrary, the drag coefficient decreases in higher Reynolds number range. The Reynolds number range in which drag reduction occurs is dependent on the cylinder diameter. As a representative case, the drag coefficient of a cylinder with d = 20 mm increases compared to that of tap water at $10^3 < Re < 3 \times 10^3$. On the other hand, drag reduction occurs at $3 \times 10^3 < Re < 3 \times 10^4$, and the maximum drag reduction ratio was approximately 55% for the 200 ppm surfactant solution at $Re = 7 \times 10^3$. In addition, the increase in drag coefficient at lower Reynolds number was caused by the existence of a wide stagnation zone around the cylinder. The increased apparent shear viscosity of the surfactant solutions is a factor in the stagnation zone.

5.4 BIOPOLYMER SOLUTIONS

Although polymer or surfactant is the effective drag reducing additives, the practical application of the drag reduction of solutions has been limited because it has an adverse effect on the environment or it makes mechanical degradation. If we notice to the mechanical degradation of solutions, the use of surfactant additives is effective for closed-loop pipeline systems. The additives are finding increasing use in city thermal energy supply pipeline systems or air-conditioning pipelines in large-scale buildings for energy conservation of pumping power at present.

As described earlier, the application of drag reduction of polymers or surfactant additives is limited to closed-loop pipeline systems, as the disposal of these solutions has an adverse effect on the environment. From the viewpoint of environmental impact, the practical application of biopolymers is convenient because drag-reducing additives biodegrade with time. Thus, if the effect of environmental loading is considered, the application to open pipeline systems can be expected. Advantages and disadvantages of drag reducing additives are summarized in Table 5.4.

Hoyt[77] reported that an aqueous solution of a polymer with a sufficiently high molecular weight exhibits a drag reduction effect, and showed that the drag reduction of sodium carrageenan polysaccharide solution, which is extracted commercial seaweed, reaches 62% in drag reduction ratio. He also showed that a polysaccharide that shows no drag reduction exists and that the drag reduction ratio of xanthan gum solutions decreases at $5000 < Re < 15{,}000$, but increases at $Re > 15{,}000$. From this, Berman[78] reported that the side chain of a polysaccharide formed from xanthan gum is in the collapsed state for the principal chain at $Re < 15{,}000$, and that the side chain is extended as a result of the turbulence of the flow at $Re > 15{,}000$.

Though, Hoyt also attempted to distinguish the drag reduction material on the basis of the conformer and structure of the chemical combination;

Table 5.4 Characteristics of drag reducing additives

Drag reducing additives (the number in parentheses shows the period of discovery)	Degradation	Environmental loading
High weight molecular polymer (1948)	×	×
Fiber or fine particle (1957)	Δ	×
Surfactant (1966)	☺	×
Biopolymer (1969)	Δ	☺

however, this seems until now unclear. He showed that the bonding configuration of the polysaccharide and structure of sugar itself is not direct cause of the drag reduction effect, and that drag reduction could occur, if it has at least to some extent a linear structure. He estimated that materials such as arabic gum and dextran, which have the advanced branching structures exhibit show no drag reduction.

Peterson et al.[79] reported that a polysaccharide extracted from yeast shows a drag reduction effect, and examined the relationship between polysaccharide concentration and drag reduction ratio. The drag reduction ratio of the polysaccharide solution increased with an increase in polysaccharide concentration. It was clarified that the drag reduction ratio is about 60% in 100 ppm polysaccharide aqueous solution.

On the other hand, Ramus et al.[80] showed that dilute aqueous solutions of polymers released by marine phytoplankton induce drag reduction phenomena in capillary pipe flow. They tested 22 species of marine phytoplankton and the products of one chlorophyte species and four rhodophyte species proved particularly effective. Gasljevic et al.[81] studied drag reducing biopolymers produced by marine microalgae for naval engineering applications. Several marine microalgal species were tested over long durations in large plate bioreactors for their production of drag reducing polysaccharides. The friction factor of a biopolymer solution from Porphyridium cruentum was measured using four pipe diameters spanning a broad range (i.e., 2, 5, 20, and 52 mm). They showed that all the drag reduction ratios measured after the retro-onset transition are independent of both velocity and pipe diameter. It gives a constant drag reduction of approximately 40%. The effects of salinity on two biopolymers of Diutan and Xanthan respectively derived from *Sphingomonas* and *Xanthomonas* bacteria were studied by Virk and Chen[82] in the high-Reynolds-number region. They reported that these biopolymers exhibit Type B drag reduction.

In relation to the biopolymer drag reduction phenomenon, Kato et al.[83] show that some kind of plant mucilage occurs Toms effect. They are Welsh onion, nameko mushrooms, Satoimo (taro), okura and sunset hibiscus (grated yan Aoi). Experiment carried out the measurement of pressure loss of two pipes with 6.3 and 6.0 mm diameter. Fig. 5.59 indicates an example of the friction factor results. It is seen that the friction factor decreases in the order of Satoimo (taro), nameko, okura and grated yan Aoi (sunset hibiscus), and the value of sunset hibiscus in no-circulation pipeline system almost agrees with the maximum drag reduction value.

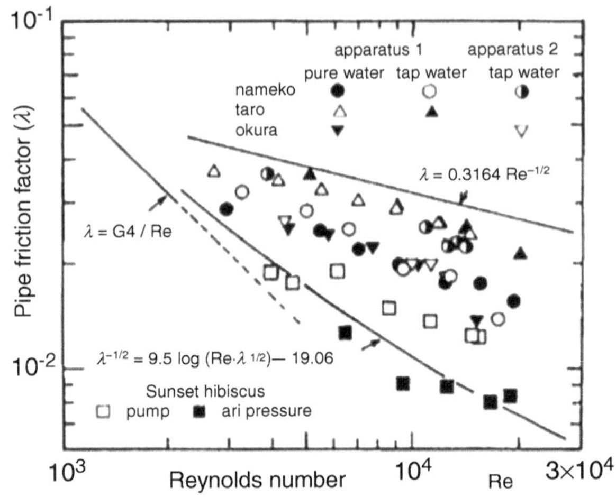

Figure 5.59 ***Friction factor of plant mucilage.***

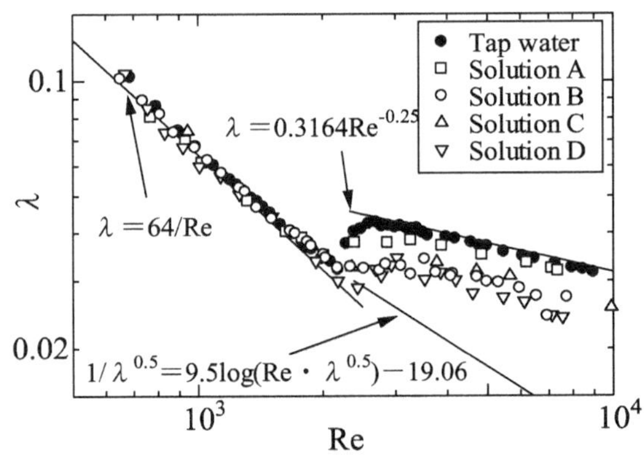

Figure 5.60 ***Friction factor of molted rice solutions.***

Watanabe and Ogata[84] clarified that dry malted rice is a turbulent-drag reducing-material from the viewpoint that it is widely used in Japan. Fig. 5.60 shows the results of the drag reduction phenomena of a culture solution of malted rice. The cultivation conditions of the test solutions are listed in the figure. Test solutions were passed through a filter, but Solution C is used only supernatant solutions. Experimental data indicate that the friction factor characteristic of culture solutions of malted rice is Type B that they become almost parallel to the Prandtl-Kármán equation. Although there is hardly the difference in the experimental data between water and culture solutions of malted rice in the laminar flow range, the friction factor decreased by about 40% in a turbulent flow range. Comparing solutions

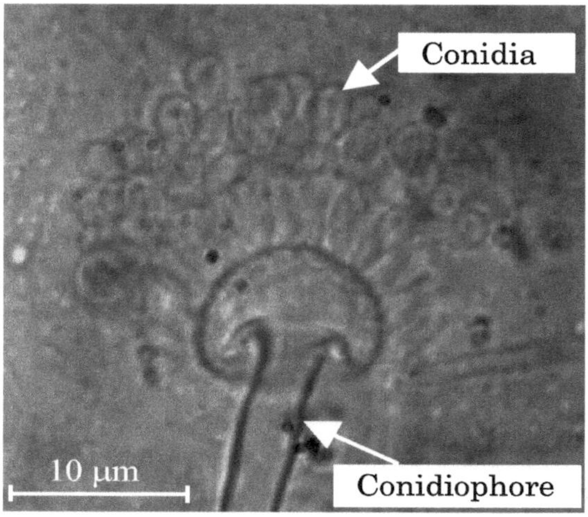

Figure 5.61 ***Micrographs of aspergillus oryzae in water.***

A and B in the figure, although the cultivation temperatures and times are the same, the drag reduction ratio of solution B is greater than that of A. The mesh sizes of the filter are 0.9 mm and 20 mm for solutions A and B, respectively. These results mean that the aspergillus itself has been removed by the small filter (0.9 mm).

Fig. 5.61 shows a micrograph of the aspergillus after cultivation. The sizes of the aspergillus including conidia and conidiospores are about 5–10 μm; therefore, the aspergillus can be clearly removed using a 0.9mm filter. The quantities of dry malted rice and water are reduced for solutions C and D, in comparison with the cases of solutions A and B. We obtained the maximum drag reduction ratio in Solution D that it is less time in the culture of rice malt. However, it is not possible to clarify the relationship between the drag reduction quantity and the culture conditions at present.

The difficulties remain in the decision for physical constants of biopolymer solutions because the biological diversity effects on the molecular structure. In other words, the same value does not have in the molecular weight even if it is the same biopolymer additives. On the other hand, the friction factor of biopolymer solutions in a circular pipe decreases roughly parallel in comparison to Kármán-Prandtl formula. Thus, Virk and Chen proposed the rearranging method for drag reduction of biopolymer solution in a circular pipe. It gives the following equation:

$$s' = \frac{1}{\sqrt{f}\big|_s} - \frac{1}{\sqrt{f}\big|_n} \tag{5.27}$$

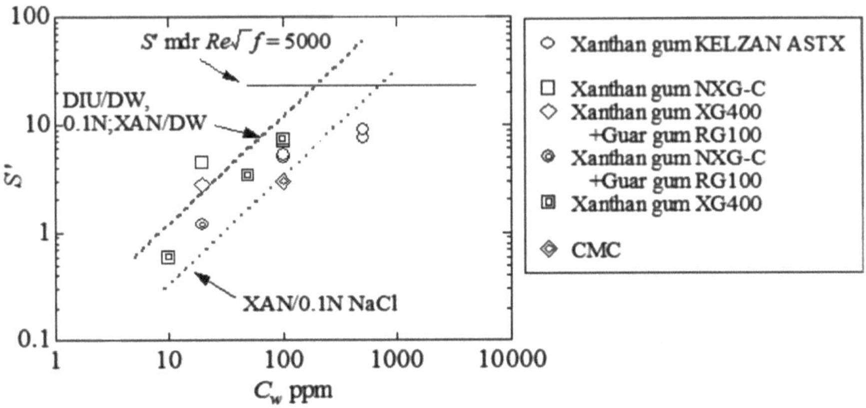

Figure 5.62 ***Flow enhancements found for biopolymer solutions.***

where *f* is the friction factor of a circular pipe and the subscripts, *s* and *n* are the mean values of biopolymer solutions and Newtonian fluids, respectively.

Fig. 5.62 indicates the experimental results of xanthan gum, gur gum, and CMC solutions. The horizontal axis indicates the concentration of solutions. Virk takes notice of the drag reduction phenomenon that the effect decreases when 0.1% NaCl is added into the solutions, and he gives the minimum value shown by the *dotted line* in the figure. Experimental data lie between two *dotted lines*, and agree well with their results. If we are going to examine the factor for drag reduction from viewpoint of continuum fluid mechanics, it may be required to measure the viscoelastic effects of the solutions. In general, it is very difficult to measure for low concentration solutions. However, we can check whether xanthan gum solutions have a viscoelastic property by a rotation of liquid in a vessel because it is very low concentration. Fig. 5.63 shows the comparison of surface profiles between distilled water and xanthan gum solutions. Evidently, the cone-shape of

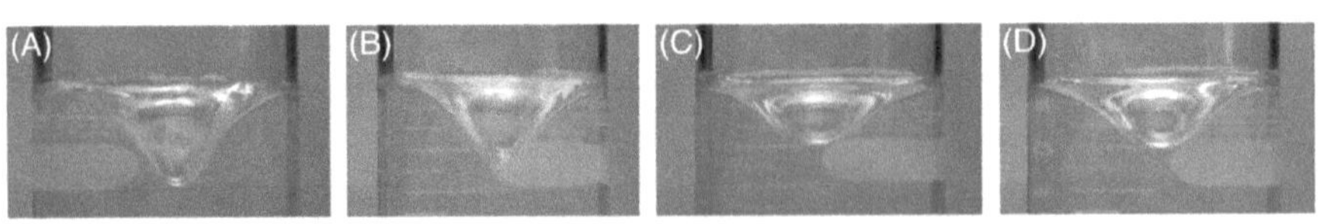

Test conditions:
- Rotating vessel: Rotating speed: 500rpm, Volume: 100cc
- Fluid: Volume: 80cc, Temp. 20°C

Figure 5.63 ***Comparison of surface profiles of rotating fluid.*** (A) Distilled water, (B) xanthan gum (10 ppm solution), (C) xanthan gum (50 ppm solution), and (D) xanthan gum (50 ppm solution).

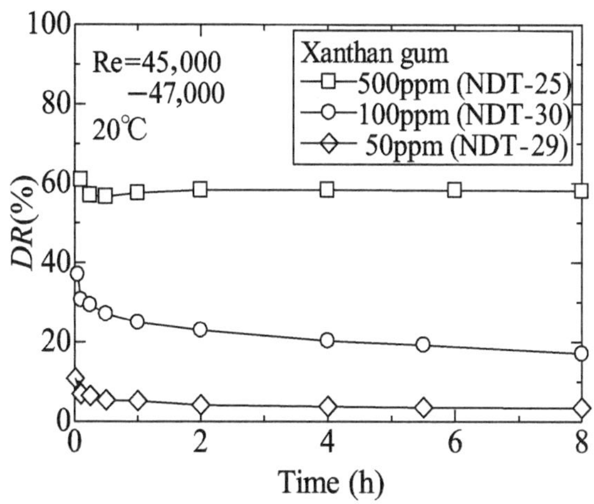

Figure 5.64 ***Evaluation of degradation of Xanthan gum solutions for pump operation time.***

depression depth of xanthan gum solutions becomes shallow in comparison to that of water. That is, viscoelastic effect occurs in it.

We focus on what matters the degradation property of the solutions. The experimental degradation data of xanthan gum solutions are shown in Fig. 5.64. Experiment carried out the measure by using a circulation pipeline system with a centrifugal pump. It is d = 15 mm in the pipe diameter. It is 25 mm in pump diameter and N = 1450rpm in the rotation speed. The vertical and the horizontal axis indicate the drag reduction ratio and the continuous running time of pump, respectively. Data show that the degradation occurs within 1 h and after it poses the constant value. It is difficult to analyze the phenomenon that it will occur by mechanical shear of a pump impeller. But, the measurement for the particle size distributions of solutions may provide a clue verifying the degradation. Although the performance of drag reduction was clarified by the degradation test, it is very difficult to know the internal structural change of the solutions. As described in the section of a pipe flow in surfactant solutions, the measurement on particle size distribution of complex mixture is a better approach to understand the phenomenon.

Fig. 5.65 shows the experimental data of the particle size distributions for xanthan gum 50, 100, and 500 ppm solutions, respectively. They were measured every hour by using a dynamic scattering photometer. Because the particle size distribution is closely allied to the scale of polymer, the change of mode in the data is a point of focus. The mode is defined as the peak of the frequency distribution. It is clearly identified as two modes in the results of xanthan gum 50 and 100 ppm solutions. In the case of xanthan

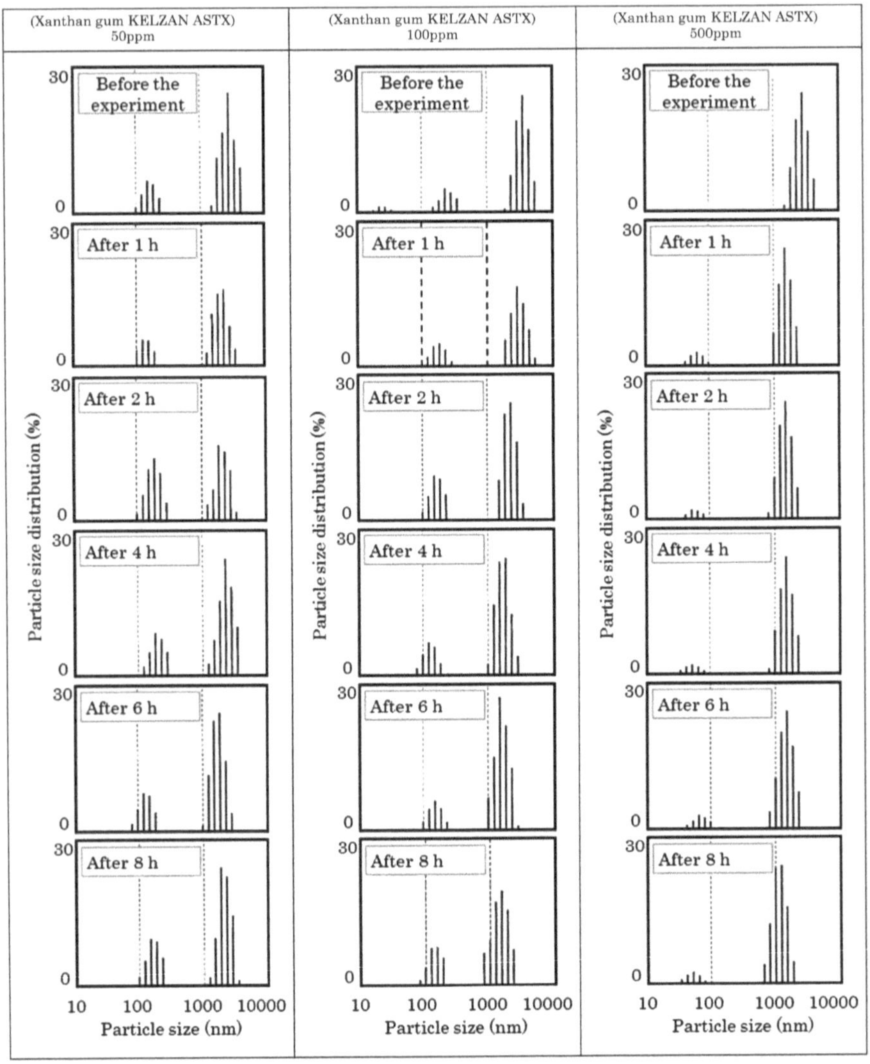

Figure 5.65 ***Comparison of particle size distributions after time course of pump operation.***

gum 500 ppm solutions, the small mode appears after 1 h of the pump running time. From Fig. 5.64, it is seen that the degradation occurs within 2 h and it becomes gentle after that. The mode variation with the time is shown in Fig. 5.66. As described earlier, the large mode decreases about by half after 2 h in each solution although the change is sufficiently small for the small mode.

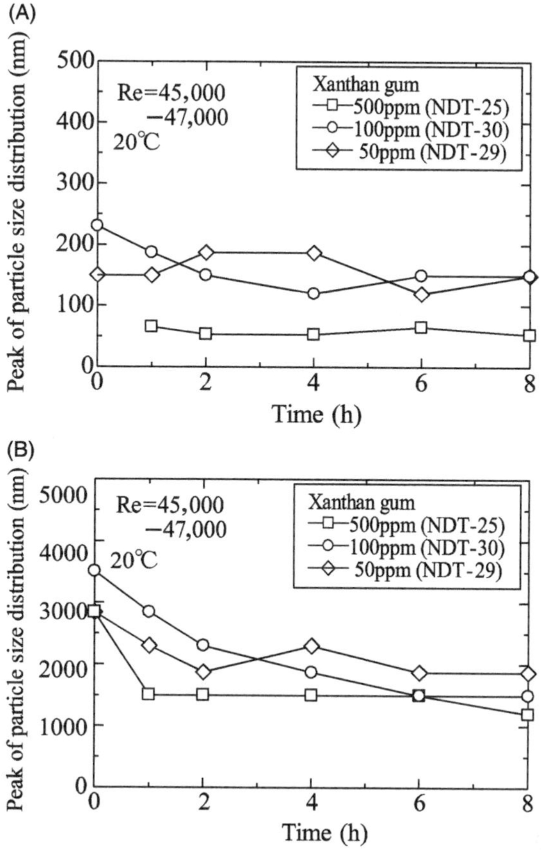

Figure 5.66 ***Variation of peak of particle size distributions.*** (A) Small mode radius, and (B) large mode radius.

On the other hand, the heat transfer performance of Xanthan gum solution may be a meaningful information to apply practically. As turbulence in the flow is modified by adding these drag-reducing additives, it is well known that an addition in surfactant solutions degrades the performance in turbulent flow range. The results of Xanthan gum solutions are shown in Fig. 5.67A and B. The *solid line* in the figures is Dittus-Boelter's experimental formula. The result of surfactant solutions is shown for comparison. As described earlier, value of Nusselt number in turbulent flow range is almost equal to that of the value of water in laminar flow rang. Obviously, the heat performance of surfactant solutions degrades in it. In the case of xanthan gum, the experimental data decrease slightly in comparison with the data of water and the decreasing ratio increases with an increase of

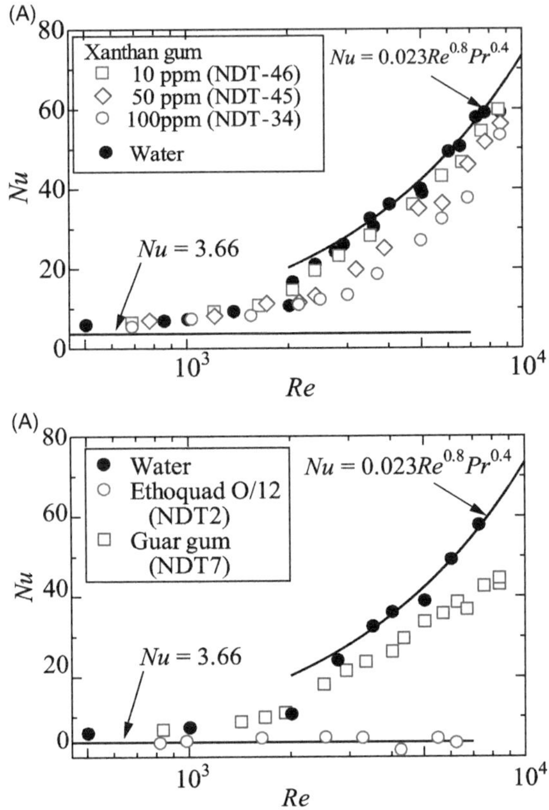

Figure 5.67 ***Nusselt number.*** (A) Xanthan gum solutions and (B) surfactant solutions.

the concentration. By the same token, it occurs in guar gum solutions. Consequently, it can be said from viewpoint of a movement for ecology that the biopolymer can expect a means for drag reducing additives to the application. Advantages and disadvantages of the drag reducing additives, which are obtained at present are summarized in Table 5.4. In future, it will be expected to provide new drag reducing additives for improving environmental problem and degradation.

Although there are still many unresolved problems in the drag reduction of a homogeneous flow, it specially lies in the transition region from laminar to turbulent flow. Much knowledge on the role of drag reduction additives may be given from the comparison between the experimental and the analytical results of the region because we can analyze the behavior of laminar flow region in many cases by using an equation motion.

REFERENCES

1. Toms BA. *Proc Int Conger Rheology* 1948;**II**:135.
2. Watanabe K. *Laminar drag reduction*. Bentham Science Pub. Co; 2015.
3. Watanabe K, Udagawa H. *AIChE J* 2001;**26**:729.
4. Virk PS, et al. *Trans ASME, J App Mech* 1970;**388**:488.
5. Navier CMLH. Mémoire sur les lois du movement des fluids. *Mémoires de l'Académie Royale des Sciences de l'Institut de France* 1823;**6**:389.
6. Watanabe K, et al. *Trans JSME* 2006;**72**(714):331 (in Japanese).
7. Quibrahim A. *Phy Fluids* 1979;**22**:784.
8. Muller-Mohnssen H, et al. *J Rheol* 1987;**31**:323.
9. Corinish RJ. *Proc. Roy. Soc. London, Ser. A* 1928;**120**:691.
10. Cohen Y, Metzner AB. *J. Rheol.* 1985;**29**:67.
11. Taylor GI. *Phil Trans Roy Soc Lond* 1923;**A223**:289.
12. Larson RG, et al. *J. Fluid Mech.* 1990;**218**:573.
13. Sadanandan B, Sureshkumar R. *Phys. of Fluids* 2002;**14**(1):41.
14. Sureshkumar R, et al. *Proc. Roy. Soc. London.* 1994;**447**:135 A.
15. Groisman A, Steinberg V. *Physical Rev. Letters* 1996;**77**(8):1480.
16. Lee SH-K, et al. *J Fluid Mech* 1995;**282**:115.
17. Watanabe K, et al. *J Visualiz* 2004;**7**(4):1.
18. Watanabe K, et al. *Trans ASME, J Fluids Eng* 2006;**128**:95.
19. Snyder HA. *J Fluid Mech* 1969;**35**(2):273.
20. Wei T, et al. *J Fluid Mech* 1991;**245**:47 19912.
21. Barcilon A, et al. *J Fluid Mech* 1979;**94**(3):453.
22. Giesekus H. *J Non-Newtonian Fluid Mech* 1982;**11**:69.
23. Von Kármán. *Z AMM* 1921;**1**(4):233 Bd, Nr.
24. Goldstein S. *Proc Camb Phil Soc* 1935;**31**.
25. Fujimoto B, Iwagaki Y. *Trans JSME* 1947;**13**(44):180 (in Japanese).
26. Ito H. *Memoirs Inst High Speed Mech* 1959;**216**:83 (Japanin Japanese).
27. Tomita Y. *Trans Bull JSME* 1961;**4**(16):671.
28. Tomita Y, Mochimaru. *Bull JSME* 1973;**16**(92):291.
29. Hoyt JW, Fabula AG. *NAVWEPS Report, No. 8635.*
30. Watanabe K. *Bull JSME* 1978;**21**(153):455.
31. Gregory N, et al. *Phil Trans Royal Soc* 1955;**248**:155 London, Ser. A.
32. Watanabe K, Kato H. Fluid machinery and fluidics. *Proc 2nd Int JSME Sym* 1972;51.
33. Bilgen E. *Trans ASME* 1971;**93**(2):85 Ser. D.
34. Schlichting H. *Boundary-layer theory*. New, York: McGraw-Hill; 1955 84.
35. Kato H, et al. *Bull JSME* 1978;**21**(161):1618.
36. Ruszczycky MA. *Nature* 1965;**206**:614.
37. White DA. *Nature* 1966;**212**:277.
38. Kato H, et al. *Bull JSME* 1972;**15**:1556.
39. White A. *Nature* 1967;**216**:994.
40. Watanabe K, et al. *Rheol Acta* 1998;**37**:328.
41. Lapple CE, Shepherd CB. *Indus Engng Chem* 1940;**32**(5):605.
42. Watanabe K, et al. *Proc ASME, FED* 1991;**110**:123 G00598.
43. Haruki N, Horibe A. *Adv Mech Eng* 2011;206219.
44. Zhang Y, et al. *J Colloid Interface Sci* 2005;**286**:696.
45. Ogata S, et al. *Proc. 4th ICEFM* 2014;1.
46. Ogata S, et al. *Trans JSME Japanese* 2013;**79**(801):879 B.
47. Virk PS. *AIChE J* 1975;**21**:625.
48. Ohlendolf D, et al. *Rheol Acta* 1997;**25**:468.
49. Watanabe K, et al. *J Visualization* 2004;**7**(4):273.

50. Imai S, Shikata T. *J Colloid Interface Sci* 2001;**244**:399.
51. Gupta VK, et al. *Phys Fluids* 2002;**14**(6):1958.
52. Arora K, et al. *Rheol Act* 2002;**41**:25.
53. Bewersdorff HW, Ohlendroff D. *Colloid Poly Sci* 1988;**266**:941.
54. Ito M, et al. *Tran JSME* 1995;**61**(590):3664 B, in Japanese.
55. Zakin JL, et al. *AIChE J* 1996;**42**:3544.
56. Warholic MD, et al. *J Fluid Mech* 1999;**388**:1.
57. Ogata S, Watanabe K. *J Fluid Mech* 2002;**457**:325.
58. Daily JW, Nece RE. *Trans ASME, J Basic Engineering* 1960;**82**:217.
59. Lin Z, et al. *J Non-Newtonian Fluid Mech* 2000;**93**:363.
60. James DF, Acosta AJ. *J Fluid Mech* 1970;**42**:269.
61. Koniuta A, et al. *J Non-Newtonian Fluid Mech* 1980;**7**:101.
62. Manero O, Mena B. *J Non-Newtonian Fluid Mech* 1981;**9**:379.
63. Usui H, et al. *J Chem Engin Japan* 1980;**13**(1):77.
64. Turgut S, et al. *J Fluid Mech* 1973;**57**:177.
65. Kim BK, Telionis DP. *Phys Fluids* 1989;**1**(2):267 A.
66. Luikov AV, et al. *Heat Trans Soviet Res* 1969;**1**(1):121.
67. Bergin C, et al. *Exper Fluid* 2001;**30**:410.
68. Ogata S, et al. *AIChE J* 2001;**30**:49.
69. Bewersdorff HW. *Proc ASME* 1996;25 FED-237.
70. Wunderlich AM, Brunn PO. *Colloid Poly Sci* 1989;**267**:627.
71. Wieselsberger C. *Ergebnisse der Aerodynamischen versuchsanstalt zu Göttingen II Lieferung.*
72. Brennen CJ. *Fluid Mech* 1970;**44**:51.
73. Bird RB, Armstrong RC, Hassenger O. *Dynamics of polymeric Liquids*. 2nd ed New York: John Wiley & Sons; 1987 2.
74. Fage A, Falkner VM. *ARC RM1369. Rep. and Mem., Aeron., (Res. Council, No. 1369).*
75. Lu B, et al. *Rheologica Acta* 1998;**37**:528.
76. Wunderlich AM, James DF. *Rheologica Acta* 1987;**26**:522.
77. Hoyt JW. *Trends Biotechnol* 1985;**3**(1):17.
78. Berman NS. Proc IUTAM Sym. *Essen/Germany* 1984;293.
79. Peterson GR, et al. *Appl Biochem Biotech* 1989;**20**(21):845.
80. Ramus J, et al. *Giotechnol Bioeng* 1989;**33**:550.
81. Gasljevic K, et al. *Proc 2nd Inter Symp Seawater Drag Reduction* 2005;559.
82. Virk PS, Chen RH. *Proc 2nd Inter Symp Seawater Drag Reduction* 2005;545.
83. Kato H, et al. *Proc Inter Con Fluid Thermal Energy Conv* 1997;**97**:395 ISSN 0854-9346.
84. Watanabe K, Ogata S. *Proceedings of the ASME IMECE2009 10421* 2009;1.

INDEX

A

Angular velocity, 116, 118, 133, 168, 177
Artificial lightweight aggregate
  dimensions of solid particles, 49
  horizontal pipes, pressure losses of, 50
Axial velocity, 41
  measured, 33
  profile, 25
    in turbulent flow range, 42, 46

B

Bend, total loss coefficient of, 47
Bingham fluids, 28, 89
Bio-fibers, 77
  dimensions, 77
  suspensions, 76–81
Biopolymer solutions, 194–202
  Aspergillus oryzae in water, micrographs, 197
  characteristics of drag reducing additives, 194
  Dittus-Boelter's experimental formula, 201
  flow enhancements, 198
  friction factor, 196
    molten rice solution, 196
    plant mucilage, 196
  Nusselt number, 202
  particle size distributions, 200, 201
  surface profiles of rotating fluid, comparison of, 198
  xanthan gum solutions, evaluation of degradation, 199
Blasius approximation, 76
Blasius' formula, 85
Boundary conditions, 20
  at wall, 21
  contact wall surface, 20
  slip boundary conditions equation, 21
  velocity profile, 33
Brownian motion, 2
Bubble
  air bubbles, tap water, 66
    application of a swirling flow, 66
    convergence radius of, 71, 72
  air bubbles, trajectory
    numerical simulation, 70
      in turbine oil, 72, 73
      in water, 71, 72
  drag coefficient, 68
  equation of motion for a air bubble
      in spiral pipe, 68
  verification of, 73
Buckingham π theorem, 7, 8, 10

C

Capillary viscometer, 31, 159, 183
Carbon black and kaolin suspensions
  fine solid particles, micrographs, 81
  flow curves, 82
Circular pipe
  bend, 45
    additional bend loss coefficient, 47
    and total loss coefficient, 47
    experimental data of, 48, 49
  blockage, 60
  Colebrook formula for, 43
  formula of laminar flow, 40
  friction factors of fly ash slurry, 92
  Navier-Stokes equations, 29
  Prandtl-Kármán formula for, 44
  pressure loss of solid-liquid two-phase flow in, 36
  solid-liquid two-phase flow , pressure loss of, 8
Circumferential velocity
  component maximum, 39
  profiles in turbulent flow range, 42
Coarse solid particle-ultrafine particle mixtures flow, 36
Colebrook formula, rough circular pipe, 43
Colloidal particle, 75
Complex mixtures flow, 4
  property, 4–12
Compressible fluid, 14
Cone and plate rheogoniometer, 111

Constant axial velocity line, 46
Continuity equation, 13
  rectangular coordinates/density of fluids, 14
Corrosion test, 165
Couette-Taylor flow, 132
Crushed stone
  dimensions of solid particles, 49
  fly ash slurry flow, 55
  horizontal pipes, pressure losses of, 50
  pressure drops of, 55
  pressure losses of horizontal pipes, 36
Curved flow path, experimental data, 48
Cylindrical coordinates, 14, 19

D

Deborah number, 132
Deformed pipe
  experimental studies, 73
  friction factor, 41
  with spiral rib/spiral wall, 37
Dodge-Metzner formula, 85
Drag of external flows, 6
Drag reducing additives, 5, 75, 194
  fiber suspensions, 76
  in turbulent flow, 75
Drag reduction ratio. *See also* Heterogeneous flow; Homogeneous flow; Pseudo-homogeneous fluid
  and the nondimensional characteristic time ratio, 109
  bio-fiber suspension under consideration, 79
  circular cylinder, 187
  vs. concentration, 79
  correlation of, 80
  defined, 113
  equation for bio-fiber suspension, 79
  vs. l/d, 79
  maximum, 56, 91
  plotted against concentration and aspect ratio, 78
  pressure coefficient of carbon black suspension, 103
  vs. pump running hour, 80
Drag reduction, classification, 3, 5
Duct, rectangular coordinates, 23

E

Euler's equation, 15, 19, 20
Experimental constants, 52
Experimental results
  correlation of, 53
  drag coefficients in surfactant solutions and tap water, 184
  for Newtonian fluids, 21, 168
  maximum drag reduction, 178
  on corrosion test, 165
  velocity profiles for surfactant solution, 171
  xanthan gum, gur gum, and CMC solutions, 198

F

Fiber suspensions, 80
Fine solid particles, 81
  physical properties, 81
Flow around rotating disk, 25
  momentum integral equation, 25
Flow behavior, 28, 75
  index, 28
  model parameter for the simulation, 5
  to set up the physical constant, 8
  using logarithmically scale, 27
Flow curve
  carbon black and kaolin suspensions, 81, 82
  fly ash slurries, 89
  graphene oxide particle suspensions, 87
  time independent fluids, 28
  xanthan gum solutions, 31
Flow fields, measurement values of, 3
Flow rate, 29, 30, 33, 94, 126, 127
Fluid
  Bingham, 89
  classification of, 1–3
  compressible, 14
  defined, 1
  density, 14
  design, 12
  incompressible, 14
  Newtonian, 17, 82, 85, 91, 104, 106, 129, 138, 142, 147, 149, 150, 167, 176, 178, 186
  non-Newtonian, 2, 20, 55, 81, 105, 141
  Oldroyd-B, 132

pseudohomogeneous, 75
rotating, 198
slip for a power-law fluid, 125
test, 128, 133, 183, 191
velocity change of an element of, 16
viscoelastic, 111, 141
viscosities, 126, 159
Fluid motion equation, 15
real fluid isothermal flow, 15
Fly ash slurries, 91
Fourier transform, 24
Friction factor, 40
biopolymer solution from porphyridium cruentum, 195
carbon black suspension, 83
categories, 44
defined, 33
drag reduction of surfactant solutions, NDT-1-NDT-5, 164
experimental data, 44
fiber suspensions based on, 76
of laminar flow, 40, 41
plant mucilage, 196
surfactant solution, Ethoquad O/12, 126
suspension in a pipe with bell mouth, 93
Friction loss, by collision, 37

G

Gas flow, 1
Gas-liquid two-phase flow, 66–73
flow characteristics, 68
Gas-solid two-phase flow, 57–66
behavior of a bubble in spiral pipe, 58
flow behavior of, 32
geothermal power generation, 66
liquid, physical constants of, 67
spiral pipe, dimensions of, 38
velocity profiles, 39
Gas-solid two-phase flow pipeline systems, 47
Görtler instability, 132
Gravitational acceleration, 52

H

Hagen-Poiseuille equation, 126
Hagen-Poiseuillie flow, 33
Heterogeneous flow. *See also* Drag reduction ratio; Gas-solid two-phase flow; Liquid-solid two-phase flow
in horizontal pipeline systems, 35
pressure drop, experimental data, 36
Heterogeneous/pseudo-homogenous fluids
drag reduction phenomenon, 5
High-speed flow, 1
Homogeneous flow, 2, 123
drag reduction. *See* Biopolymer solutions; *See also* Polymer solutions; Surfactant solutions
Horizontal pipeline systems
pressure drop experimental data, 41
Horizontal pipes, pressure losses, 36
Hydraulic transport system, 3, 36, 37
coarse solid particle-ultrafine particle mixtures flow, 55
fly ash slurries, used as carrier fluids, 56
heterogeneous flow, 35
minimum value used as a target for, 51
pipeline systems, 37
solid-liquid two-phase flow, 36

I

Incompressible fluid, 14, 15
Inlet length/pitch, swirling number variation, 43
Inviscid fluid flows vorticity, 19

K

Kaolin suspensions, flow curves, 82
Kármán' s formula, 142, 149
Kinetic energy, 132

L

Laminar region, velocity profiles, 25, 32
Laplace operator, 18
Laser-induced-fluorescence (LIF) flow visualization, 133

Lift acting on a sphere, measurement for, 114–121
calculation results
lift coefficient, 120
lift condition, 119
lift force, 118
particle trajectory, 120
sphere, in stationary condition, 121
with no-lift condition, 118
dimensions of test spheres, 115
equilibrium radios, 116
experimental apparatus, 115
flow around a sphere in a rotating vessel, 117
motion equation for a sphere, 114
particle trajectory
and dynamic equilibrium, 117
Liquid-solid particle suspensions, 81–82
flow in a divergent channel, 94–101
comparison of pressure profiles, 95
flow visualization, 98
pressure recovery coefficient, 96
carbon black suspension, 96
reduction ratio for, 97, 100
separation location, asymmetric diffuser, 99
two dimension flow visualization system, 94
flow in a pipe, 82–94
drag reduction ratios, 88
flow curve, 83
fly ash slurries, 89
graphene oxide particle suspensions, 87
friction factors, 83, 88, 92
carbon black suspension, 83
defined, 85
fly ash slurry, 92
Kaolin suspension, 83
laminar flow, 85
SiO_2 suspension, 83
spiral pipe for Newtonian fluids, 89, 90
suspension in a pipe with bell mouth, 93
graphene oxide particles and the suspensions, 86
II-type spiral pipe, 91
semitheoretical equation for friction factor, 82
shapes of pipe entrance, 93
flow over a cylinder and sphere, 101–114
characteristic time, T_h and relaxation time, 109, 110
dimensions of test sphere, 103
drag coefficient, 104
fine solid particle suspensions, 108
sphere in Newtonian range, 107
sphere in tap water for, 108
drag reduction ratio, 103, 113
equation of motion considering the Basset term, 104
Newtonian fluid, 104
experimental apparatus for falling sphere test, 103
fall velocity
sphere in fine solid particle suspension, 106
sphere in water, 105
first normal stress difference, 112
pressure coefficient of circular cylinder, 102
separation angle for carbon black suspension, 102
thrust and the shear rate, 111
Liquid-solid two-phase flow
axial velocity profiles, 39, 42
calculated sectional area, 40
circumferential velocity profiles, 42
comparison with experimental data, 39, 42
constant axial velocity line, 42
correlation of experimental results, 52
drag reduction ratio, 51
experimental data and analytical results, 52
flow behavior, 40
in spiral pipe, 52
of polyethylene resin pellets mixtures, 53
flow streaks in spiral pipe, 38
outline of spiral pipes, 38
particle concentration distribution, 54

particle motion in pipe, 37
pressure drops of crushed stone, 55
pressure loss and flow patterns, 36
spiral pipe bends, 45
total bend loss coefficient, 48
total loss coefficient of bend, 47
variation of swirling number, 43
velocity profiles in laminar region, 39
Liquid/gas flows, drag reduction technique, 8

M

Micro PIV system, 129
Mixed matter, on pressure loss, 4
Multiphase flow, 2

N

Navier-Stokes equations, 17–19, 25, 29
incompressible fluid with constant viscosity, 18
vector formula, 19
Navier's hypothesis, 21, 125
Newton's law of viscosity, 1
Newton's second law, 15
Newtonian fluid, 1, 17, 22, 27, 28, 40, 82
flow, 20, 21
friction, 87
in rectangular coordinates, 17
Navier-Stokes equation, 22–27
spiral pipe, 39
Newtonian single-phase flow
in turbulent flow range, 42
Non-Newtonian fluid, 1, 20, 27, 91
for carrier fluids, 55
rheological equation, 27
Non-Newtonian viscosity, 28, 82, 183
Nonlinearity, in viscous behavior, 75

O

Oleyl bis (hydroxyethyl) methylammonium chloride, 159
Oleyl dihydroxyethyl amine oxide (ODEAO), 157
Oleyltris (hydroxyethyl) methylammonium acetate, 157

P

Parallel beam, 54, 133
Particle concentration distribution, 54
Particle size
distributions, 81, 86, 146, 159, 161
heterogeneous mixtures flow, 2
methods for obtaining, 5
typical example of, 3
Photo-transistor, 54
Physical constants of mixed matter, 6
Physical quantity, correlation chart of, 6
Pipe flow, pressure loss, 5
Piston, 31
Polyethylene resin pellets mixtures
flow behaviors of, 53
Polymer solutions, 22, 124
flow around a rotating disk, 148
analytical and experimental results, comparison between, 145
experimental apparatus, 142
overall view/torque pick-up detail, 142
flow angle
defined as, 148
enclosed rotating disk, 151
rotating disk in a large tank, 150
moment coefficient
high concentration polymer solutions, 145
of newtonian fluids, 143
rough surface disk in Newtonian fluids, 147
rough surface disk in Peo 10 solutions, 147
separan solutions/Peo solutions, 144
non-Newtonian viscosity effect, 151
path line of a rotating flow, 149
power-law models, analytical results, 145
rough surface disk, 146
streaks of flow on a rotating surface, 148
value of relative roughness, 146
flow between two coaxial cylinders, 130
experimental apparatus
for flow visualization system, 133
flow pattern, 131

Polymer solutions (*Cont.*)
- formation process
  - taylor volrecies in tap water, 135
- Gõrtler wavelength of polymer solutions, 137
- measurement, using Snyder's formula, 135
- Taylor cell of separan 100 ppm solutions
  - forming process of, 135
- Taylor vortex
  - apparent viscosity of Separan 100 solutions, 138
  - early stage, formation of polymer solutions, 136
  - effects of viscoelasticity, 137
  - in gap of rotating cylinders, 131
- flow in a pipe, 124
  - flow rate increasing ratio, 127
  - friction factor
    - equation, 125
    - of PEO solutions in a pipe with, 124
    - surfactant solution, 126
  - image of drag reduction phenomenon, 125
  - power-law Reynolds number, 125
  - velocity profile in
    - a micro-channel, 130
- flow past a sphere, 152
  - drag coefficient
    - in high Reynolds number range, 156
    - of sphere, 155
  - drag reduction ratio and Weisenberg number, relationship, 156
  - fall velocity, 153
    - ratio and diameter ratio in tap water, relationship between, 154
    - ratio of velocities in, 155
  - flow visualization results
    - for flow around a fall sphere, 158
  - onset of drag reduction
    - of Separan solutions, 153

Polypropylene pellet, horizontal pipes, pressure losses of, 50

Power law exponent, 28

Power-law fluids, 27, 28
- physical constant, determination of, 29

Power-law index, 28

Prandtl-Kármán formula, 44, 87
- for circular pipe, 197

Pressure drop, experimental data, 49

Pressure loss formula, 52

Pressure transducer, 31, 182
- differential, 76

Pseudo-homogeneous fluid, 75
- drag reduction in, 75. *See also* Biofiber, suspensions; Lift acting on a sphere, measurement for; Liquid-solid particle suspensions

R

Reynolds number, 34, 41, 69, 76, 78, 90, 93, 96
- range, 44

Rheological equation, 156

Rotating disk, flow, 139

Rotational cone viscometer, 111

Runge-Kutta method, 104, 116

S

Sectional area, calculation, 40

Shear stress, 21, 25, 29, 76, 77, 141
- bio-fiber suspensions, 76

Shear thinning phenomena effect, 128

Single phase flow, 1, 2

Slip velocity
- velocity profile, 33
- water-repellent wall, 22

Slurry, 81
- spiral tube to transport, 89

Solid particles
- dimensions of, 49
- drag reduction ratio
  - variation for delivered mass fraction, 51

Solid-liquid mixture, pressure loss, 49

Solid-liquid two-phase flow, 36
- pressure loss and flow patterns, 36

Spherical coordinates, 14, 19

Spiral pipe
- assumed velocity profiles water, 39
- bends, 45

crushed stone-fly ash slurry flow, 55
dimensions of, 38
drag reduction, 51
liquid-solid two-phase flow, 52
flow behavior, 53
flow streaks, 38
friction factor, 41, 44
outline of, 38
pressure loss, 41
types, 38
Spiral rib/spiral wall
deformed pipe, 37
Spiral ribbed pipe
sand-water mixtures flowing, 37
Steady laminar flow, in rectangular duct, 22
Stresses element of fluids, 18
Surfactant solutions, 157
flow around a rotating disk, 169–179
comparison of parameters, 178
differential equation solved for, 173
flow angle of, 176
flow model of velocity profiles
in boundary layer on a rotating disk, 170
flow patterns near rotating disk, 175
moment acting on one side, 173
moment coefficient, 174
acting on one side, 173
for Newtonian fluid, 178
in a dilute polymer solution, 178
in surfactant solutions, 177
nondimensional velocity profile, for surfactant solutions, 171
radial and tangential components of shearing stress, 172
Reynolds number range, 178
velocity profiles with no-slip boundary condition, 172
flow between two coaxial cylinders, 166–169
formation process of Taylor cells, 166, 167
Görtler wavelength of surfactant solutions, 168
flow in a pipe, 157–166
drag reduction of, 164
surfactant solutions, NDT-1-NDT-5, 164
surfactant solutions, NDT-5-NDT-8, 165
effect of additives on viscosity of Ethoquard O/12, 161
experimental results on corrosion test, 165
test solutions, 159
particle size distributions, 161–163
flow past a circular cylinder, 179–193
bubble formation and complicated rheological behavior, 180
drag coefficients of surfactant solutions, 185
drag reduction ratio of a circular cylinder, 187
effect of *(l/d)* on drag coefficient, 186
effect of *Um/d* on drag reduction ratio, 193
experimental apparatus for, 181, 182
flow around a circular cylinder, 188
for drag reducing polymer solutions, 180
numerical simulation results
for flow around a circular cylinder, 190
pressure coefficient, 183
pressure profiles of a circular cylinder, 191
shear induced state (SIS), 179
Strouhal number, 180
velocity profiles in the wake of a cylinder with, 190
viscosity of surfactant solutions, 183
Suspension, 75
biofiber, 76
liquid-solid particle, 81
Swirling flow, 38, 42

T

Taylor expansion, 15
Taylor number, 132, 133, 137
Taylor vortex, 131, 134, 135
Taylor-Couette flow, 132
Test solid particle, dimensions, 49

Tetradecyl trimethyl ammonium bromide, 180
Tomita's formula, 142
Toms effect, 5
Total bend loss coefficient, 48
Turbulent flow, 5
 pipe flow, pressure loss, 44
 range
  axial velocity profiles, 42
  Blasius formula, 49
  circumferential velocity profiles, 42
  formula of a Newtonian single-phase flow in, 85
  reduction of ratio in, 125
  value of Nusselt number in, 201
  velocity profiles, 41

V

Velocity change, of element of fluids, 16
Velocity profile
 and pressure distribution, 6
 anion polyacrylamide solutions, 129
 boundary layer, 25
 circumference component of, 39, 42
 in flow direction of elliptic pipe, 89
 in micro channel, 129
 logarithmic, 170, 179
 nondimensional, 171
 of rayon and hardwood-kraft suspensions, 76
 one-seventh -power law, 42
 polymer solutions in a micro-channel, 130
 rectangular duct, 25
 water in the spiral pipe, 68
 with fluid slip, 32, 33
  determination of, 22
Viscosity, 15, 18, 29, 76, 109, 126
 apparent, 28, 85
 of surfactant solutions, 183
 relative, 55
Visualization dye, 133

W

Water flow, 1
Water-repellent wall
 friction factor of PEO solutions, 124
 slip velocity, 22
Waved pipe wall, pressure losses, 38
Weissenberg number, 155, 156

X

Xanthan gum solutions. *See also* Biopolymer solutions
 drag reduction ratio, 194
 flow curve of, 31
 in collapsed state, 194

Y

Yang's modules, 5, 6
Yield stress, 28
Young's modulus, 76–78

FSC
www.fsc.org
MIX
Papier | Fördert gute Waldnutzung
FSC® C083411

Kolibri 360 GmbH
Ferdinand-Jühlke-Straße 7
99095 Erfurt, Deutschland
produktsicherheit@kolibri360.de